WORD PROBLEMS
from *Literature*

Help Students Master Problem Solving
in Elementary to Middle School Math

Books by Denise Gaskins

Let's Play Math: How Families Can
Learn Math Together—and Enjoy It

Math You Can Play Series
Counting & Number Bonds
Addition & Subtraction
Math You Can Play Combo
Multiplication & Fractions
Prealgebra & Geometry

Playful Math Singles Series
Let's Play Math Sampler
70+ Things To Do with a Hundred Chart:
Word Problems from Literature
Word Problems Student Workbook
312 Things To Do with a Math Journal

The Adventurous Student Journals Series

Printable Activity Guides for Teachers and Homeschoolers

A PLAYFUL MATH SINGLE

WORD PROBLEMS

from *Literature*

Help Students Master Problem Solving in Elementary to Middle School Math

SECOND EDITION

Denise Gaskins

Tabletop Academy Press

Tabletop Academy Press, Boody, IL, USA
tabletopacademy.net

ISBN: 978-1-892083-64-7
Print version 2.0

Contents

Preface to the Second Edition

As a math coach, I love showing adults and children how to look at math with fresh eyes, to explore the adventure of learning math as mental play, which is the essence of creative problem-solving. Mathematics is not just rules and rote memory. Math itself is a game, playing with ideas.

I've written several books to help families play math together. But at heart, I've always been a fiction fan—especially fantasy fiction. And this book, *Word Problems from Literature*, lets me bring that love of story to the surface.

This is one of my all-time favorite books, and I've had so much fun with this new edition: adding stories, writing make-your-own-problem prompts, sneaking extra teaching tips into the worked-out solutions, creating an almost-magical guide to helping kids reason their way through math problems.

To provide support when your children get stuck, I added my favorite problem-solving tip, the Four Questions that push students to apply their own common sense, emphasizing the importance of reasoning about math. And to give you a boost when *you* get stuck, I beefed up the explanations of the advanced puzzles, adding several tips on understanding and teaching fraction division and other monster calculations.

There's a new section in the student workbook on "What If I'm Not Good at Math?" to help children develop a problem-solving mindset. Plus, I wrote more than 30 new word problem prompts to get your students writing their own creative math stories.

Most of all, I've tried as much as possible to encourage both adults and students to work at making sense of the math, seeing how the numbers relate, avoiding the crutch of standard textbook rules so you can experience the joy of figuring things out for yourself.

Those of us who made it through school math by memorizing and following rules eventually paid the price. We came to the point where our minds could hold no more, where the rules we had learned all jumbled together, where we felt lost in the dark as the rock began to crumble beneath our metaphorical feet while the wizard rammed his staff into the ground and cried, "You shall not pass!"

Some people reach the point of mental exhaustion with upper-elementary fractions or middle school ratios and percents, while others make it into algebra or geometry before they crash. A rare few have a good enough mental filing system that they can proceed as far as calculus before it all falls apart.

Estimates vary, but anywhere from half to three-fourths of adults suffer from some level of math anxiety due to their school experience.

Can we spare our children this fate?

We must strive to teach math in a way that makes sense, where children don't just memorize the rules we tell them but see clearly how math concepts connect, drawing their own conclusions, building their understanding into a solid foundation for future learning.

And we must give our students the tools to build on this foundation, problem-solving tools that will help them face and conquer any new math monster that comes their way.

This book will help you do that.

Have fun playing math with your kids!

—DENISE GASKINS,
RURAL ILLINOIS, JULY 2022

Storying—
encountering the world
and understanding it contextually
by shaping ideas,
facts,
experience itself
into stories—
is one of the most fundamental means
of making meaning.
As such, it is an activity
that pervades all learning.

—GORDON WELLS

Word Problems as Mental Manipulatives

It is the duty
of all teachers,
and of teachers of mathematics
in particular,
to expose their students
to problems
much more than to facts.

—PAUL HALMOS

1

Puzzles Build Mathematical Thinking

ARCHIMEDES TRIED TO FIND THE distance around a circle and almost discovered calculus. Pierre de Fermat predicted the result of a gambling game and laid the foundations of probability. Leonhard Euler went for an afternoon walk over the bridges of Königsberg and invented topology. Georg Cantor created a way to count infinity and opened a whole new world of modern math.

Through the centuries, mathematics has grown as mathematicians struggled with and solved challenging puzzles.

Problems are the raw material of math, the ore we dig, grind up, and melt, refining it to produce ideas. Our understanding of math grows as we play with problems, puzzle them out, and look for connections to other situations. The threads that connect these problems become the web of ideas we call mathematics. Each puzzle we solve adds a new thread to the web, or strengthens one that already exists, or both.

If we want our children to learn real math, we need to offer them plenty of problems to solve. A child may work through several pages of number calculations by rote, following memorized steps, but a good problem demands more thought.

The story in a word problem puts flesh on the abstract bones of arithmetic, encouraging children to ponder what it means for one thing to be bigger than another, or smaller, or faster, or slower, or made up of several parts.

Math professor Herb Gross says: "We teachers so often hear students summarize a course by saying, 'I could do everything except the word problems.' Sadly, in the textbook of life, there are only word problems."

Our children will meet numbers in many guises throughout their lives. Few of these will be as straight-forward as a textbook word problem, but real-life problems and school math stories will always have certain things in common. Quantities will be related to each other in a given proportion. Situations will be complex, and solutions may require many steps.

Story problems give students a chance to grapple with these complexities in a controlled environment, where we can increase the difficulty in stages over several years.

The Purpose of Word Problems

Word problems feed a student's imagination. Like other puzzles, word problems are often artificial, but that needn't diminish our pleasure in solving them.

In working a puzzle, we expect to find difficulties and setbacks. That's part of the game. Similarly, when students approach word problems as puzzles, they become less concerned with rushing to an answer and more interested in figuring out the relationships within the story.

Playing with puzzles strengthens mathematical understanding, according to retired professor Andre Toom. Originally from Russia, Toom taught math in several colleges and universities around the world.

Here's how he explains the purpose of story problems:

"Word problems are very valuable—not only to master mathematics, but also for general development.

"Especially valuable are word problems solved with minimal scholarship, without algebra, even sometimes without arithmetic, just by plain common sense. The more naive and ingenuous is the solution, the more it provides the child contact with abstract reality and independence from authority, the more independent and creative thinker the child becomes.

"When we teach children to solve problems in school, we do not expect them to meet exactly and literally the same problems in later life. Mathematical education would be next to useless if its only use were literal. We want much more, we want to teach children to solve problems in general.

"In this respect traditional word problems are especially valuable, because to solve a word problem, you have to understand what is said there.

"This function of word problems is very poorly understood. The main educative value of word problems is that they serve as mental manipulatives, paving children's road to abstract thinking.

"For example, coins, nuts, and buttons are clearly distinct and countable, and for this reason are convenient to represent relations between whole numbers. The youngest children need some real, tangible tokens, while older ones can imagine them, which is a further step of intellectual development. That is why coin problems are so appropriate in elementary school.

"Pumps and other mechanical appliances are easy to imagine working at a constant rate. Problems involving rate and speed should be common already in middle school. Trains, cars, and ships are so widely used in textbooks not because all students are expected to go into the transportation business, but for another, much more sound reason: These objects are easy to imagine moving at constant speeds.

"There is an important similarity between children's play and mathematics: In both cases, creative imagination is essential."

—Andre Toom

The Trouble with Word Problems

How can we teach our students to reason their way through math problems? We must help them develop the ability to translate real-world situations into mathematical language.

Most young children solve story problems by the flash-of-insight method, hearing the problem and knowing by instinct how to solve it. This is fine for simple problems like "Four kittens played with a yarn ball. Two more kittens came to join the fun. Then how many kittens were playing with the yarn ball?"

When problems grow more difficult, however, that flash of insight becomes less reliable. We find our children fidgeting with their paper, staring out the window, complaining, "I don't know what to do. It's too hard."

They need a tool that will work when insight fails.

In solving a textbook word problem, students blaze a trail through the unknown. They must:

♦ Read the problem and understand what it's asking.

♦ Translate the problem into a mathematical calculation or algebraic equation.

♦ Do the calculation or solve the equation.

♦ Make sense of the answer, explaining how it relates to the original problem.

The first two steps give students the most trouble. They could calculate just fine, if they could decide which calculation to do.

But they don't know how to translate the problem from English into Mathish.

One common classroom approach emphasizes *key* or *signal* words. For instance, we can tell our children that a problem asking "How many more?" will probably require subtraction. The question asks for the difference between two quantities, and *difference* is the answer when you subtract. But this technique only works for the simplest word problems.

Unfortunately, key words can mislead a careless reader.

For example, consider this question: "What must we add to 2 to get 7 as the sum?" The words *add* and *sum* are designed to lead an unwary child into the trap of answering 2 + 7 = 9.

Or this one: "There are 21 girls in a class. There are 3 times as many girls as boys. How many boys are in the class?" Would the word *times* trick your child into multiplying 21 × 3 = 63?

I do teach a few key words to my students. My favorite is the translation "*of* = multiply" when dealing with fraction and percent problems. But I want my kids to read a math problem and to analyze what is happening, no matter what words are used to describe the situation. For this, they need a more powerful tool.

Four Common-Sense Questions To Solve Any Problem

Because all people begin their school career by solving problems intuitively, children expect to look at a math puzzle and instantly see the answer. As students progress through the years, their math problems develop layers of complexity. Solution by intuitive leap becomes impossible. Too often a frustrated child concludes, "I'm just not good at math."

But the truth is that *nobody* is good at math, if you define "good at math" to mean they can see the answer instantly. Here's a more useful definition: You're good at math if you have problem-solving tools and know how to use them.

And *that* is something everyone can learn.

When children are faced with a math problem, they need to combine the given facts in some way to reach the required answer. But rarely can they do it in a single leap. So encourage them to take one little step at a time. Even if they can't solve the problem, can they think of a way to get closer to the goal?

Teach your child to ask these four questions…

(1) What Do I Know?

Read the problem carefully. Reread it until you can describe the situation in your own words.

List the facts or information given in the problem. Notice math vocabulary words like *factor, multiple, area,* or *perimeter.* What do you remember about those topics?

Sometimes a problem tries to trick you. Watch out for mixed units: If one length is given in inches and another is given in yards, make them consistent.

Try to express the facts in math symbols or using the visual algebra of a bar model diagram.

(2) What Do I Want?

Describe the goal, what the problem is asking you to find. What will your answer look like?

Notice important words like *product, sum, next,* or *not.* Small words like "not" are especially easy to miss.

Try to express the goal in math symbols or using the visual algebra of a bar model diagram.

(3) What Can I Do?

Imagine yourself in the story situation, applying your hard-earned common sense. If this actually happened to you, what would you do?

Mix things around in your mind. Combine the given facts.

Have you worked a problem like this before? How did you solve that one? Will that method, or something like it, work here?

If you're using a bar model, think about ways you might move or cut the blocks to discover new relationships.

Try a tool from your problem-solving toolbox:

♦ Draw and label a diagram or sketch.

♦ Act the problem out, step by step.

♦ Make a systematic list, chart, table, or graph.

♦ Look for a pattern.

♦ Simplify the problem. Try it with smaller numbers.

♦ Change your focus. Restate the problem in another way.

♦ Look for a related problem. How is it the same? How is it different?

♦ Think about "before" and "after" situations.

♦ Work backwards. Start at the end of the problem and find a path back to the beginning.

♦ Guess and check. Try something to see if it works, and then make sure you know why.

If you are completely stumped, explain the problem to another person. Talking it over might help unclog your brain, opening your eyes to clues you had missed until you put them into words. Or take a break to let the problem simmer in your unconscious mind while you do other things.

(4) Does It Make Sense?

Don't neglect this last step! When you think you have found the answer, always look back at the original problem one more time.

Does your answer make sense? Did you leave anything out?

Can you think of a way to confirm that your answer is right?

Can you think of another way you might have gotten the answer? If you see an alternate approach, would that method have been easier? Make a note of any ideas you come up with. You may need them to solve your next puzzle.

Sometimes, instead of working through a whole page of math problems, take time to look intently at one problem. See how many different ways you can find to solve it. Such an exercise builds flexible thinking, showing students the depth of their own knowledge, helping them reason more creatively the next time they get stuck.

Bar Model Diagrams as Pictorial Manipulatives

One of my favorite tools for translating elementary and middle school word problems into math is a *bar model diagram*. The process of drawing and working through a bar model leads the child step by step through all four problem-solving questions.

Bar models are a pictorial form of algebra in which we represent quantities (both known and unknown) as block-like rectangles. The child imagines moving these blocks around or cutting them into smaller pieces to find a useful relationship between the known and unknown quantities. In this way, the abstract mystery of the word problem becomes a shape puzzle: How can we fit these blocks together?

For children who have played with Legos or Cuisenaire rods, this visual approach can reveal the underlying structure of a word problem, which helps them see how it's like others they've already solved, just as detectives search for a criminal's *modus operandi* (MO). Recognizing a math problem's structure helps kids solve the case.

It's a trick well worth learning, no matter what math program you use.

All bar model diagrams (also called *Singapore math models* or *measuring tape diagrams*) descend from one basic principle: The whole is the sum of its parts. If you know the value of both parts, you can add them to get the whole. If you know the whole total and one part, you can subtract what you know to find the other part.

Whole

Part A	Part B

The most basic bar diagram: Two parts make a whole.

Recall the problem designed to stump a child using key words: "What must we add to 2 to get 7 as the sum?" We can draw a rectangular bar to represent the total amount. Then we divide it into two parts, representing the number we know and the unknown part. Now it's easy to see the answer:

7

2	?

Seven is the sum, the whole thing. What must be the missing part?

As the number relationships in math problems grow more complex, the bar model may be split into more than two parts. Also, the parts may relate to each other in ways that require a more elaborate diagram. Multiplication, division, or fraction problems will involve several parts that are the same size, called *units*. But even with a complicated story, the solution begins by drawing a simple bar to represent one whole thing.

For instance: "There are 21 girls in a class. There are 3 times as many girls as boys. How many boys are in the class?" To show three times as many girls as boys, we start with a bar for the

Boys [?]

Girls [| |]

21

The bar model helps children see that they need to
divide, not multiply, to find the number of boys.

number of boys. That's one unit, and then we need three units to
show the number of girls.

Teaching Tips

I think the best way to learn to solve story problems is by *not solving the problems.* Look at the story and show how to think about it, without being distracted by calculations. Work on learning to apply common sense.

When I teach my students to draw bar diagrams, I do it apart from their daily homework. My kids may work their daily homework by whatever method they choose—including doing it in their heads and just writing an answer—as long as they can explain the logic of their solution. But in our story problem journal, they have to draw the bars.

As compensation for the extra pencil work of drawing, we don't calculate the answer. Once they show me how to set up the bars and tell me the steps for solving it, they are done with that problem. They think they are getting off easy, because they skipped the actual multiplication or subtraction or whatever. But I want them to focus on reasoning about the problem situation, learning how the bar diagram tool works before they get to the tough problems where they need it.

We begin by drawing long rectangles—imagine Lego blocks or Cuisenaire rods. I say, "Let's imagine the books/fish/snowballs set out in a row." We write labels beside each bar, to identify its meaning in the story. We build the diagram as we discuss the problem, adding numbers inside a block, using brackets to group the bars together or to indicate a specific section of a bar.

If your student has trouble figuring out where the numbers go in the diagram, you might ask, "Which is the big amount, the whole thing? What are the parts? How do the parts relate to each other?"

To solve a bar model diagram, your student must learn two simple but important rules:

The Whole Is the Sum of Its Parts

Bar diagrams rely on the inverse relationship between addition and subtraction: The whole is the sum of its parts. No matter how complicated the word problem, the solution begins by identifying a whole thing made up of parts.

Simplify to a Single Unknown

You cannot solve for two unknown numbers at once. You must use the facts given in your problem and manipulate the blocks in your drawing until you can connect one unknown unit (or a group of same-size units) to a number. Once you find that single unknown unit, the other quantities will fall into place.

Practice bar models with your child until drawing a diagram becomes almost automatic. Start with simple story problems that are easy enough to solve with a flash of insight. Discuss how you can show the relationships between quantities, translating the English of the stories into a bar model picture. Then work up to more challenging problems.

The following chapters provide a range of story puzzles from early-elementary problems to middle-school stumpers. Let's get started …

Lay the Foundation:
One-Step Problems

Word Problems Inspired by
Mr. Popper's Penguins

*A family of four adopts several penguins
and teaches them to perform tricks.*

IN THIS CHAPTER, I'LL DEMONSTRATE the bar model problem-solving tool in action with a series of early-elementary word problems. For your reading pleasure, I've written stories that might happen in a situation like the family-favorite read-aloud book, *Mr. Popper's Penguins* by Richard and Florence Atwater.

Try your hand at the problems before reading my solutions. Can you draw a bar model diagram to represent each story situation?

[1]

During the winter, Papa read 34 books about Antarctica. Then he read 5 books about penguins. How many books did Papa read in all?

[2]

When Papa opened the windows and let snow come into the living room, his children made snowballs. The girl made 18 snowballs. Her brother made 14 more than she did. How many snowballs did the boy make? How many snowballs did the children make altogether?

[3]

Papa had 78 fish. The penguins ate 40 of them. How many fish did Papa have left?

[4]

The family dressed in their best clothes for their meeting with the show manager. Mama had a ribbon 90 centimeters long. She had 35 cm left after making a bow for her daughter's hair. How much ribbon did Mama use to make the bow?

[5]

The penguins did theater shows for 2 weeks. They performed 4 shows every week. How many shows did the penguins perform?

[6]

While they were staying at the hotel, Papa put a leash on one of the penguins and took him for a walk. They climbed up 3 flights of stairs. There were 10 steps in each flight.
Then the penguin flopped onto his stomach and slid down all the stairs. He pulled Papa with him all the way. How many steps did Papa fall down?

Bar Models and Answers

[1]

During the winter, Papa read 34 books about Antarctica.
Then he read 5 books about penguins. How many books did
Papa read in all?

One of the best ways to approach a word problem is to ignore the numbers. Encourage your child to think about what is happening and describe the story in general words, without using numbers: "Papa read some books. He read a lot of books about Antarctica. And he read some books about penguins, too."

Whenever you or your child feel confused about a word problem, come back to this basic starting point. Tell the story without numbers, so you can focus on the relationships. After you make sense of what is happening, then you can put the numbers back in.

We can imagine the books standing together on a library shelf, sorted by topic. We draw one long bar to represent all the books. Then we split it into two parts: books about Antarctica, and books about penguins.

Books [_____ | _____]

Notice that the diagram is not to scale. That's the algebraic power of the bar model: The bars are variables, and we needn't worry about getting a perfect proportion. As long as we fit our model to the story relationships, the numbers will work themselves out.

The story asks us to find the total number of books, so we indicate this with a bracket either above or below the bar. We put a question mark for our goal.

Books

?

Now we've drawn out the basic relationship in our word problem, so we're ready to add in the numbers. Each number goes inside the bar that represents that amount.

Books

34	5

?

The whole is the sum of its parts, so we need to add to find our answer.

$$34 + 5 = 39$$

Papa read thirty-nine books.

[2]

> *When Papa opened the windows and let snow come into the living room, his children made snowballs. The girl made 18 snowballs. Her brother made 14 more than she did. How many snowballs did the boy make? How many snowballs did the children make altogether?*

In early arithmetic, most addition and subtraction problems involve active situations: joining, separating, growing, shrinking, giving, and taking away. The movement in these stories offers children a strong hint about how to do the math.

But when problems are static, without movement, they pose a greater challenge. Stories about collections with different categories of items, groups with different types of people or animals, or comparisons between quantities of different sizes—none of these situations give students an action clue that shows what to do.

With young children first learning to deal with static situations, I like to offer them an action they can visualize. I might say something like:

"Pretend that we lined up all the snowballs side-by-side, with the boy's snowballs in one row and his sister's snowballs in another. Which line would be longer? How many of the snowballs would sit side-by-side? How many would be left without a partner? How many more snowballs would the girl have to make if she wanted the rows to match?"

To model a comparison, we draw a bar for each thing we're comparing. The left edges of the bars line up, making it easy to see which is longer or shorter.

Here our bars will represent the number of snowballs each child made. The boy made more snowballs than his sister, so his bar needs to be longer.

Sis [   ]

Bro [   ]

The question has two parts. We need to figure out how many snowballs the boy made, and the total of all the snowballs together. Add brackets and question marks to show the quantities we need to find.

Sis [   ] ⎤
                    ⎬ ?
Bro [   ] ⎦

⎣_____⎦
        ?

That gives us the basic situation of our word problem. Now we are ready to fill in the number clues. It's easy to write 18 in the bar for the girl's snowballs, but how can we show that her brother made fourteen more than she did?

When two blocks are the same length, we often draw dotted lines to connect their endpoints. We mark off a section of the brother's bar to make a block equal to his sister's, then write 14 into the extra chunk.

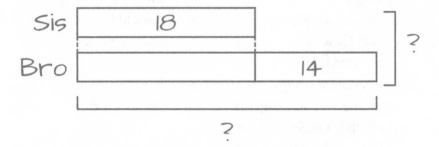

The whole is the sum of its parts, so we need to add the two chunks to find out the boy's number.

$$18 + 14 = 32$$

The boy made thirty-two snowballs. And now we can add both children's amounts together to find the total.

$$18 + 32 = 50$$

The children made fifty snowballs altogether.

[3]

Papa had 78 fish. The penguins ate 40 of them. How many fish did Papa have left?

As a math textbook moves into working with larger numbers, it often returns to problems with a simpler structure. Here we have an action problem, easy for children to visualize. The whole bar is all the fish, with two parts: fish eaten, and fish that remain.

Fish

Now to add the details—the clues from our word problem. We have two numbers. One tells the whole amount, so we'll show that with a bracket. The other number is one part, and our goal is to find the missing part.

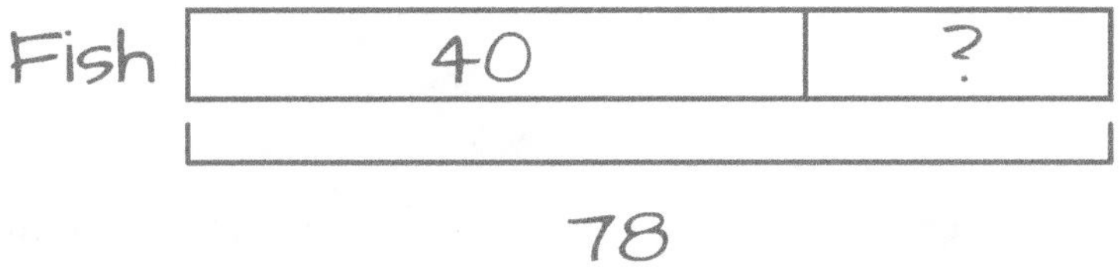

Fish | 40 | ?

78

Adults typically consider this a subtraction problem. But many children prefer to think in terms of addition: "How much do I add to go from forty up to seventy-eight?"

$$40 + 38 = 78$$
$$78 - 40 = 38$$

Whichever way we think it through, there were thirty-eight fish left.

[4]

The family dressed in their best clothes for their meeting with the show manager. Mama had a ribbon 90 centimeters long. She had 35 cm left after making a bow for her daughter's hair. How much ribbon did Mama use to make the bow?

Students who try to reason directly from the story to the calculation may get confused by the structure of this word problem. The number thirty-five seems to be the answer after a subtraction, but how can we subtract a number we don't know?

The bar diagram shows the connection between this problem and the previous one. We have a single bar (the ribbon) made of two parts: the bow and the extra.

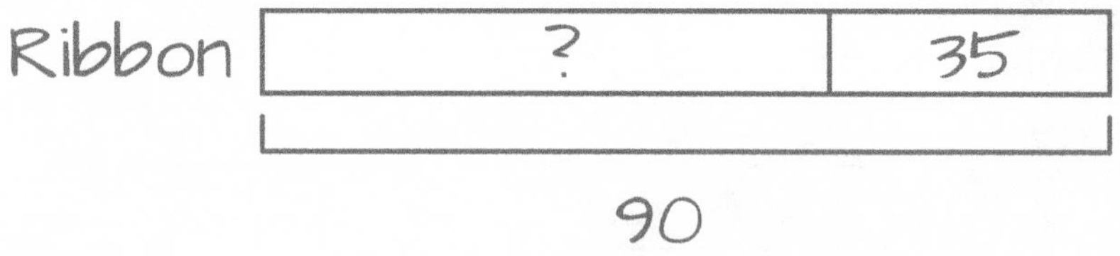

Ribbon | ? | 35 |

90

Once again, the student may calculate the answer by counting up or by using subtraction.

$$35 + 55 = 90$$
$$90 - 35 = 55$$

Either way, it took fifty-five centimeters of ribbon to make the bow.

[5]

The penguins did theater shows for 2 weeks. They performed 4 shows every week. How many shows did the penguins perform?

This problem introduces a new type of number: a *this-per-that* quantity, also known as a *rate*. The penguins performed four shows per week.

In most problems, a this-per-that relationship will require multiplication or division. How can the student know whether to multiply or divide? In simple problems, it should be easy to see which calculation to use, so we'll come back to this topic at a deeper level in later chapters.

We can draw a single bar for the number of penguin performances. Split the bar in half to represent two weeks.

In a bar diagram, two or more parts the same size are called *units*. Here each unit is one week's worth of theater shows.

Shows | 4 | 4 |

?

This early in their school career, most students will not even notice the multiplication. They will simply add the units.

$$4 + 4 = 8$$
$$2 \times 4 = 8$$

The penguins performed eight shows in all.

[6]

> *While they were staying at the hotel, Papa put a leash on one of the penguins and took him for a walk. They climbed up 3 flights of stairs. There were 10 steps in each flight.*
>
> *Then the penguin flopped onto his stomach and slid down all the stairs. He pulled Papa with him all the way. How many steps did Papa fall down?*

Here we have another this-per-that quantity: ten steps per flight of stairs. Each flight of stairs is one unit. When there are same-size units in a diagram, we don't have to repeat the number. Just write it on the first unit.

Unit = 10

3 units = 3 × 10 = 30

Papa fell down thirty steps in all.

Make Your Own Math

One of my favorite math journaling activities is to have children write and share their own problems. We'll talk about this more in Chapter 7, but for now you can try the following story problem prompts.

What kind of things do the characters in your story world count? Do they measure or cut things, or do crafts? What do they like to collect or share, or what are their favorite foods? What kind of math stories will you create?

[7]

Write a changing-amount problem set in the world of a book or movie you enjoy. Your story will include:

- A beginning amount.
- Some type of change—joining or separation, giving or taking, growth or shrinking.
- And the final amount.

Tell any two of these numbers. Then ask your reader to figure out the third.

[8]

Write a collection problem set in the world of a book or movie you enjoy. Your story will include:

- A whole amount (the collection).
- Two types or groups of people, animals, or things that are parts of the collection.

Leave either the whole or one of the parts a secret. Then ask your reader to find it out.

[9]

Write a comparison problem set in the world of a book or movie you enjoy. Your story will include:

- A smaller amount.
- A larger amount.
- The difference between them.

Leave one of these numbers a secret. Then ask your reader to figure it out.

[10]

Write a problem that contains some sort of this-per-that unit.

For example, where in your story world would people think about the number of legs per animal, cookies per child, dollars per item they buy, or something like that?

Your problem has three numbers:

♦ How many same-size units.

♦ The size of a single unit.

♦ The total amount.

Tell any two of these numbers. Then ask your reader to figure out the third.

[11]

Write any kind of problem you like.

A Note about Copyright and Trademarks

Most of the books and movies mentioned in this book are the protected intellectual property of their authors or estates, or of the company which bought those rights.

I have created math problems inspired by these stories, but I've peopled my work with generic characters (a father, a faun, a rogue starship captain) to avoid infringing on the original author's creation.

When you and your students write problems for your own private use, feel free to use your favorite characters from any story. This is a form of fan fiction. But if you decide to share your creation beyond your own home or classroom, then be sure to "genericize" it first. Change or remove the proper names, using general descriptions instead.

For example, if your children love the Star Wars movies, they might want to use a Jedi in their story problem. Instead, encourage them to write about "an interstellar justice warrior with an energy sword." Or "an alien master of martial arts training a cocky but inexperienced young apprentice."

We'd love to add your child's story to the Student Math Makers Gallery.[†]

[†] *tabletopacademy.net/math-makers*

Build Modeling Skills:
Multistep Problems

3

Word Problems with Ben Franklin

A young boy grows to adulthood in colonial America.

As OUR CHILDREN MOVE INTO middle-elementary math, the numbers get bigger, and the problems get more complex. Many word problems will take more than one step to solve. This time, I've created stories in the colonial world of Ben Franklin, inspired by the history book *Poor Richard* by James Daugherty.

Try your hand at the problems before reading my solutions. Can you draw a bar model diagram to represent each story situation?

[1: 1706]

> *Ben helped his father make 650 tallow candles. After selling some, they had 39 candles left. How many candles did they sell?*

[2: 1718]

> *Ben sold 830 newspapers. His brother James sold 177 fewer newspapers than Ben.*

(a) How many newspapers did James sell?

(b) How many newspapers did they sell altogether?

[3: 1724]

Ben loved to visit the London book shops. In one shop, he found 6 shelves of books on geography, travel, and maps from around the world. Each shelf held the same number of books. There were 30 books altogether. How many books were on each shelf?

[4: 1726]

Ben and his friends made a club called "The Junto" to read books and discuss ideas. Ben read 7 science books. He read 5 times as many history books as science books. How many more history books than science books did he read?

[5: 1732]

Ben collected donations for many worthy organizations. He had 2,467 pounds in a bank account to start a new hospital. A friend gave him another 133 pounds. How much more money must Ben collect if he needs 3,000 pounds for the hospital?

[6: 1776–1785]

While in France to negotiate a treaty, Ben went to a fancy party. There were 1,930 women at the party. There were 859 fewer men than women. How many people were at the party altogether?

Bar Models and Answers

[1]

Ben helped his father make 650 tallow candles. After selling some, they had 39 candles left. How many candles did they sell?

Here is another of those confusing "subtract an unknown number" problems, like the ribbon in the last chapter. We need a single bar divided into two parts.

With the diagram, we can see clearly which calculation we need.

$$650 - 39 = 611$$

Ben and his father sold 611 candles.

[2]

Ben sold 830 newspapers. His brother James sold 177 fewer newspapers than Ben.
    (a) How many newspapers did James sell?
    (b) How many newspapers did they sell altogether?

In early two-step problems, the textbook asks explicitly for each step. Later, students will have to figure out which intermediate steps they need to find the answer. For this problem, the main danger is that your child will insert a mental period after "James sold 177" and not notice the word "fewer."

Whenever your child feels confused about a word problem, try covering up the numbers with removable sticky notes. Talk about the story without numbers, so you can focus on the relationships between quantities. After your child makes sense of what is happening, then you can put the numbers back in.

This is a comparison problem. We need a bar for the newspapers Ben sold and another for the papers James sold. Because James sold fewer, his bar will be shorter.

Add question marks to identify the quantities we want to find. To show that we are looking for the total, we put a bracket to the right of our drawing, encompassing both bars.

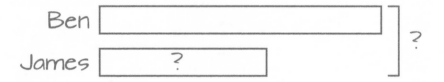

Now it's time to put in the number clues. And here's something new: 177 is not the size of either bar, but the difference between them. Oh, wait, we have seen that before—remember the snowballs? We can mark off a section of Ben's bar to make a block equal to James's, then write 177 into the extra chunk.

The bar model leads us through the calculations:

$$830 - 177 = 653$$
$$830 + 653 = 1483$$

James sold 653 newspapers, and the brothers sold 1,483 altogether.

[3]

> *Ben loved to visit the London book shops. In one shop, he found 6 shelves of books on geography, travel, and maps from around the world. There were 30 books altogether. How many books were on each shelf?*

The words *each*, *every*, and *average* often signal a this-per-that (or rate) problem. Here, we need to figure out how many books per shelf. This-per-that problems require multiplication or division. Since our problem starts with the total number of books and shares them evenly among the shelves, we will need division.

Multiplication and division are inverse operations. Just as the basic "part + part = whole" diagram showed the inverse relationship between addition and subtraction, so now students will use the "(number of units) × (size of units) = total" diagram to represent both multiplication and division problems.

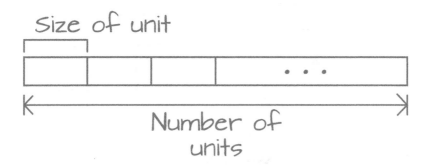

The this-per-that rate is the size of the unit. In this problem, it's the number of books per shelf.

Now we can add the number clue and a question mark showing our goal. Remember, with identical units we only write a number in the first one.

$$6 \text{ units} = 30$$
$$1 \text{ unit} = 30 \div 6 = 5$$

There were five books per shelf.

[4]

Ben and his friends made a club called "The Junto" to read books and discuss ideas. Ben read 7 science books. He read 5 times as many history books as science books. How many more history books than science books did he read?

Here we have another comparison problem. The "5 times as many" shows that we're dealing with same-size units of books: seven books per unit.

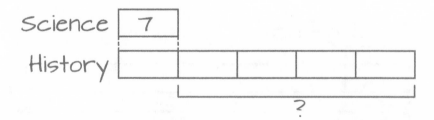

Most people solve this problem by multiplying to find the number of history books and then subtracting to figure out "how many more?" But our bar diagram contains enough information to lead to the answer in a single step.

$$1 \text{ unit} = 7$$
$$4 \text{ units} = 4 \times 7 = 28$$

Ben read twenty-eight more history books than science books.

[5]

Ben collected donations for many worthy organizations. He had 2,467 pounds in a bank account to start a new hospital. A friend gave him another 133 pounds. How much more money must Ben collect if he needs 3,000 pounds for the hospital?

This problem requires an algebraic way of thinking. How can we find the unknown number in a calculation like this:

$$2467 + 133 + ? = 3000$$

The bar diagram helps us see the number relationships.

It's a "whole is the sum of its parts" problem, but the known part is itself made up of two pieces. We need to add those pieces before we can subtract to find out the size of the missing part.

$$2467 + 133 = 2600$$
$$3000 - 2600 = 400$$

Or we could subtract each of the known parts individually.

$$3000 - 2467 = 533$$
$$533 - 133 = 400$$

Either way, Ben still needs to collect four hundred pounds.

[6]

While in France to negotiate a treaty, Ben went to a fancy party. There were 1,930 women at the party. There were

859 fewer men than women. How many people were at the party altogether?

This is like the newspaper problem earlier, but with bigger numbers. Notice that the math book is no longer asking a question for each step in the problem. Careless students will do one calculation and then move on to the next problem. Remind your children to check whether they've answered the question before going on.

Again, the main danger is that your student will read the sentence as "There were 859 men" without noticing the rest of the phrase.

Our bars will represent the men and women, as if they were standing in lines. There are fewer men, so we make the women's bar longer. And we're trying to find out how many people altogether. So here is the basic bar diagram.

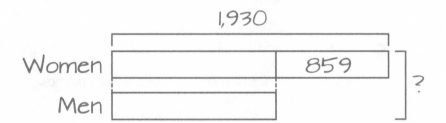

Now we can add in the number clues: the number of women, and the difference between the women and the men.

Most people will first subtract to find out how many men were there, and then add to get the total number of people.

$$1930 - 859 = 1071$$
$$1930 + 1071 = 3001$$

There were 3,001 people at the party.

But the bar diagram makes me think of another way to reason through the problem. What if there had been the same number of men as women? It's not hard to double a number. Of course, then I'd have to subtract the extra, imaginary men. But in this case, that's an easy calculation.

$$2 \times 1930 = 3860$$
$$3860 - 859 = 3001$$

There's never only one way to solve a math problem. Don't be surprised if your children think of some strange, new way to figure out an answer. As long as their solution makes sense, it's correct.

Make Your Own Math

Think about the characters in your favorite story world. What do they do during the day? Do they cook, or travel, or build things? What kinds of things would they use math to figure out? Here are a few story problem prompts your children may enjoy.

[7]

Write a changing-amount problem set in the world of a book or movie you enjoy. Your story will include:

♦ A beginning amount.

♦ Some type of change—joining or separation, giving or taking, growth or shrinking.

♦ And the final amount.

Tell numbers for the final amount and *either* the initial amount or the change that happened. Ask your reader to find the number you leave out.

[8]

Write a comparison problem set in the world of a book or movie you enjoy. Your story needs four numbers:

♦ A smaller amount.

♦ A larger amount.

♦ The difference between them.

♦ The total of both amounts.

Leave two of these numbers a secret. Then ask your reader to figure them out.

[9]

Write a problem about a collection, set, or group composed of three or four parts.

Tell the total amount and the sizes of all but one of the parts.

But for some of the parts, don't give away the number. Just tell a clue—for example, how much more or less than another part. This makes a multistep puzzle-within-a-problem.

Can your reader figure it out?

[10]

Write a problem that contains some sort of this-per-that unit.

For example, where in your story world would people think about the number of ounces per recipe, apples per basket, cost per book, or something like that?

Your problem has three numbers:

♦ How many same-size units.

♦ The size of a single unit.

♦ The total amount.

Tell any two of these numbers. Then ask your reader to figure out the third.

[11]

Write any kind of problem you like.

Master the Technique:
From Multiplication
to Fractions

Word Problems Inspired by Narnia

Four children discover a magical world with talking animals.

IN UPPER-ELEMENTARY MATH, PROBLEMS TAKE a large step up on the difficulty scale. Students are more mature and can read and follow more complex stories. Multistep word problems become the new norm, and proportional this-per-that relationships show up often. Over time, fractions grow to be a dominant theme.

For your reading pleasure, I've written stories as they might happen in a world like that of C. S. Lewis's *The Lion, the Witch and the Wardrobe*.

See how many bar model diagrams you can draw before reading my solutions.

[1]

> *35 fauns came to a midnight party of fauns and dryads in the forest. There were 3 times as many dryads as fauns. How many creatures were at the party? If the whole group split equally into 5 large circles for dancing, how many were in each circle?*

[2]

> *A studious professor had 486 books at his house, some in the library room and some in his study. There were 50 books more in the library than in the professor's study. How many books were in the study?*

[3]

> *The witch queen had 300 wolves and dwarfs as servants at her house, which was really a small castle. There were 10 more wolves than red dwarfs. The number of red dwarfs was twice the number of black dwarfs. How many black dwarfs worked at the witch's house?*

[4]

> *The beaver's wife baked a large marmalade roll for dessert. She cut 1/6 of the roll for her and her husband to share, and then she sliced up 4/6 of the roll for the children. What fraction of the marmalade roll was left?*

[5]

> *The beavers had a pitcher of milk. They poured 1/2 of it into glasses for the children to drink with dinner. Then they poured 1/8 of the pitcher into their cups of after-dinner tea. How much of the pitcher of milk did the beavers use?*

[6]

> *The witch queen's sledge got stuck in the mud and slush 24 times before she gave up and decided to walk. 2/3 of those times, the witch made her prisoner get out and help push. How many times did the prisoner have to push the sledge?*

[7]

⅖ of the creatures waiting with the magical lion prince at his pavilion were dryads and naiads. There were 20 dryads and naiads in all. How many creatures were waiting with the prince at his pavilion?

[8]

The lion prince sent 20 of the swiftest creatures to follow the wolf and rescue the prisoner. ⅖ of these creatures were eagles, griffins, and other flying fighters. The rest were centaurs, leopards, and other fast-running beasts. How many of these creatures could not fly?

[9]

The witch queen's evil minions used 4⅖ meters of rope to bind the lion prince's legs together. They used ³⁄₁₀ m less of rope to tie him tightly to the stone. How many meters of rope did the horrid creatures use in all?

[10]

⅘ of the sea people who sang and played music for the coronation party were mermen. The rest were mermaids. If there were 8 mermaids, how many sea people performed at the party?

Bar Models and Answers

[1]

> *35 fauns came to a midnight party of fauns and dryads in the forest. There were 3 times as many dryads as fauns. How many creatures were at the party?*
>
> *If the whole group split equally into 5 large circles for dancing, how many were in each circle?*

Careless students will multiply three times thirty-five, but a bar diagram shows their error. We have fauns and dryads, so let's imagine them standing in two lines, ready to dance. The fauns make a single unit, while the dryads have three times as many.

Fauns | 35 |
Dryads | | | |
?

So we have four units in all.

$$1 \text{ unit} = 35$$
$$4 \text{ units} = 4 \times 35 = 140$$

There were 140 creatures at the party.

This question had two parts, with the second question introduced by the word "if." This sometimes confuses children, who wonder, "But what if it was different?" In a math problem, *if* statements do not offer options. They merely provide information and lead to the question. Students who find the *if* confusing may ignore the word, or even cross it out.

Now we need to imagine the whole group split into five danc-ing circles. This is the basic "(number of units) × (size of units) = total" diagram for multiplication and division problems.

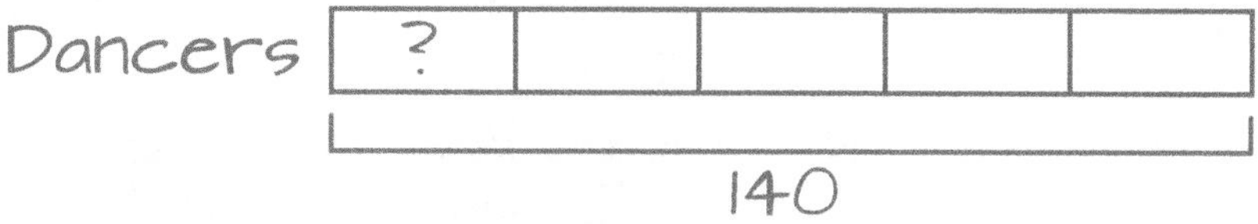

Dancers | ? | | | | |

140

$$140 \div 5 = 28$$

There were twenty-eight dancers in each circle.

[2]

A studious professor had 486 books at his house, some in the library room and some in his study. There were 50 books more in the library than in the professor's study. How many books were in the study?

In my experience, students who have not learned to think in bar diagrams will divide the books in half and then subtract fifty from that number to get their answer. A bar diagram shows the correct path to a solution. The library has more books, so it gets the longer bar. Filling in the number clues, we get:

Library | | 50 |
Study | ? |
486

Dividing by two only works if each amount is exactly half of the total—that is, if the bars are the same length. First, we need to subtract the fifty extra books. Then we can divide the rest of the books between the two rooms.

$$486 - 50 = 436$$
$$436 \div 2 = 218$$

The professor had 218 books in his study.

[3]

> *The witch queen had 300 wolves and dwarfs as servants at her house, which was really a small castle. There were 10 more wolves than red dwarfs. The number of red dwarfs was twice the number of black dwarfs. How many black dwarfs worked at the witch's house?*

Here we have a comparison problem, but it's more complicated than the ones we've seen before. We must build up the bar models bit by bit. Let's start with just the names. And since the red dwarfs are compared to both the other groups, let's put them in the middle.

Wolves

Red Dwarfs

Black Dwarfs

With a complex problem, I skip around and do the easiest part first. "Twice the number" is a familiar diagram, so let's start by drawing bars for the dwarfs. The question mark shows our goal: How many black dwarfs?

Wolves

Red Dwarfs

Black Dwarfs ?

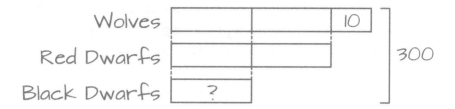

Next we add the wolves, which match the red dwarfs plus ten extra. And a bracket to the right shows the total number of servants.

Now we can see we have five of our unknown unit (which is the number of black dwarfs), plus ten more wolves, to make a total of three hundred creatures serving the witch.

$$5 \text{ units} + 10 = 300$$
$$5 \text{ units} = 300 - 10 = 290$$
$$1 \text{ unit} = 290 \div 5 = 58$$

There were fifty-eight black dwarfs.

[4]

The beaver's wife baked a large marmalade roll for dessert. She cut ⅙ of the roll for her and her husband to share, and then she sliced up ⁴⁄₆ of the roll for the children. What fraction of the marmalade roll was left?

In word problems, children first experience fractions not as abstract numbers, but as operators. That is, they find "half of" or "⅔ of" something. The fraction has meaning not in itself but in relationship to something else, some quantity we are viewing as the whole thing.

Notice the parallel between *groups of* (multiplication) and *part of* (fractions). Conceptually, fractions are much more closely related to multiplication and division than to counting or addition. As students make the effort to draw bar models, they begin to see this connection and to build an intuition about fractions that will serve them well when problems grow more challenging.

We begin with a single long bar for the gloriously sticky roll.

Roll

We divide the bar into equal-size units to make the fractions. The beaver's wife is slicing the roll into sixths.

Roll

Now we are ready to represent our number clues. We can color the beavers' one-sixth of the roll, plus the children's four-sixths. That makes a total of five-sixths of the roll that gets eaten.

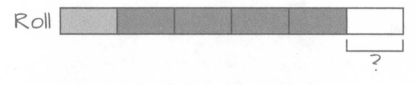

$$\tfrac{1}{6} + \tfrac{4}{6} = \tfrac{5}{6}$$
$$1 - \tfrac{5}{6} = \tfrac{1}{6}$$

So one-sixth of the marmalade roll remains.

[5]

The beavers had a pitcher of milk. They poured ½ of it into glasses for the children to drink with dinner. Then they poured ⅛ of the pitcher into their cups of after-dinner tea. How much of the pitcher of milk did the beavers use?

The fractions in this problem are a little more difficult than the last one, having different (but related) denominators. The first part of the story is easy to draw. A bar for the pitcher of milk, with half of it gone.

Milk

But how can we show the extra eighth of the pitcher that gets poured into the tea? We must go back and subdivide the bar so that each half becomes four-eighths. This is the bar diagram equivalent of finding a common denominator for fractions.

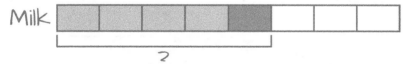

Milk

As we create the equivalent fraction, can you see why the numerator goes up in direct proportion with the denominator? As the size of the pieces gets smaller, the number of pieces increases *by the same factor.* That's why the rule for equivalent fractions tells you to multiply both top and bottom by the same number.

Milk

?

Now we can mark the extra one-eighth of the milk poured out for the tea.

$$\tfrac{1}{2} + \tfrac{1}{8} = \tfrac{4}{8} + \tfrac{1}{8} = \tfrac{5}{8}$$

They used five-eighths of the pitcher of milk.

[6]

> *The witch queen's sledge got stuck in the mud and slush
> 24 times before she gave up and decided to walk. ⅔ of those
> times, the witch made her prisoner get out and help push.
> How many times did the prisoner have to push the sledge?*

Bar diagrams are almost like magic for fraction (and later, percent) problems, because the "whole" bar can represent any amount. The process of cutting the whole quantity into fractional pieces clearly relates to division, which is itself closely related to multiplication, and seeing the multiple units of the bar model helps children reason about this connection.

You may have heard the keyword slogan "*of* means multiply," but it's actually more useful when you think about it backwards: "multiply means *of*." Children (and adults) have no idea how to visualize an abstract expression like "⅔ × 24," but "⅔ of 24" makes perfect sense.

We draw a bar for the number of times the sledge got stuck, and then split it into thirds. Two of the units represent the times that the prisoner helped push them out of the mud.

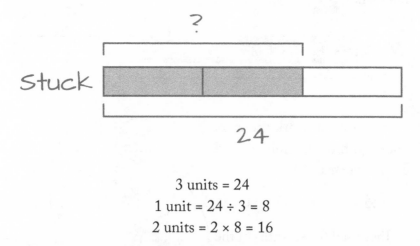

$$3 \text{ units} = 24$$
$$1 \text{ unit} = 24 \div 3 = 8$$
$$2 \text{ units} = 2 \times 8 = 16$$

The prisoner pushed the sledge sixteen times.

[7]

⅖ of the creatures waiting with the magical lion prince at his pavilion were dryads and naiads. There were 20 dryads and naiads in all. How many creatures were waiting with the prince at his pavilion?

Now we begin to reap the full benefit of our work on learning bar diagrams. Instead of struggling to understand the algebraic reasoning required to solve something like "⅖ × ? = 20," our students can draw a picture.

Part of something is 20. How much is the whole thing?

Imagine the creatures at the pavilion lined up in a row: one bar. Split the bar into fifths, and then two units will represent the dryads and naiads. We want to figure out how many creatures are in the whole bar.

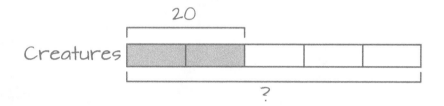

$$2 \text{ units} = 20$$
$$1 \text{ unit} = 20 \div 2 = 10$$
$$5 \text{ units} = 5 \times 10 = 50$$

There were fifty creatures with the lion prince at his pavilion.

[8]

The lion prince sent 20 of the swiftest creatures to follow the wolf and rescue the prisoner. ⅖ of these creatures were eagles, griffins, and other flying fighters. The rest were centaurs, leopards, and other fast-running beasts. How many of these creatures could not fly?

This problem uses the same numbers and the same basic bar model diagram as the last problem. But students must read and understand the story to know how the numbers relate to the diagram. Also, they need to beware how simple, easy-to-miss words like "not" can change their answer.

In this problem, our bar model represents the creatures sent to rescue the witch's prisoner. We divide the bar into fifths, so two units represent the flying fighters. We're interested in the creatures that can't fly.

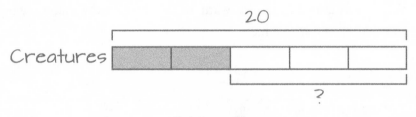

$$5 \text{ units} = 20$$
$$1 \text{ unit} = 20 \div 5 = 4$$
$$3 \text{ units} = 3 \times 4 = 12$$

So twelve of the creatures could not fly.

[9]

> *The witch queen's evil minions used 4⅖ meters of rope to bind the lion prince's legs together. They used ³⁄₁₀ m less of rope to tie him tightly to the stone. How many meters of rope did the horrid creatures use in all?*

Measurements lead naturally to mixed-number problems. Here again, we have different (but related) denominators—this time, fifths and tenths.

Notice that we deal with fractional measurements before we start working with decimals. Decimal numbers are a special (and less intuitive) form of fractions. Children need plenty of practice manipulating basic fractions to form the conceptual foundation for understanding decimals.

Start by drawing one bar that is five units long for the first rope. Each unit will represent one meter, and the last unit divides into fifths to show the mixed number. The second rope is shorter. A question mark shows we want to find the total length of rope, but first we will have to figure out how long the second piece was.

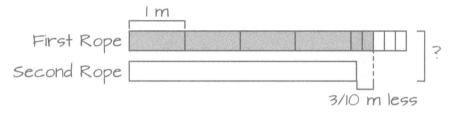

Because the fractions are related, we can convert the fifths into tenths. Split each fifth in half: $\frac{1}{5} = \frac{2}{10}$. Then we can measure the second rope.

$$4\tfrac{2}{5} - \tfrac{3}{10} = 4\tfrac{4}{10} - \tfrac{3}{10} = 4\tfrac{1}{10}$$

Now we are ready to add both pieces of rope together.

$$4\tfrac{4}{10} + 4\tfrac{1}{10} = 8\tfrac{5}{10} = 8\tfrac{1}{2}$$

They used eight and one-half meters of rope.

When you put your fraction answers into simplest form, that's a polite way to help your reader visualize what the number means.

[10]

> *⅕ of the sea people who sang and played music for the*
> *coronation party were mermen. The rest were mermaids.*
> *If there were 8 mermaids, how many sea people performed at*
> *the party?*

After the previous problem, this one may seem almost easy. Yet this question is an early experience with the ultimate challenge of fraction division. Written as an abstract equation, this problem asks:

$$8 \div \tfrac{1}{5} = ?$$

…and that's another way of saying:

$$\tfrac{1}{5} \times ? = 8$$

$$\text{or } \tfrac{1}{5} \text{ of } what = 8?$$

Students don't need to recognize this connection between fraction multiplication and division yet, but it's good for us adults to be able to see it.

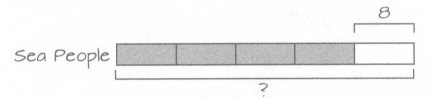

We can model this problem with one bar for the sea people, divided into fifths. Four units are mermen, so the last unit must be the eight mermaids. We want to know the whole quantity.

$$1 \text{ unit} = 8$$
$$5 \text{ units} = 5 \times 8 = 40$$

There were forty sea people performing at the coronation party.

Make Your Own Math

Have you had fun writing story problems of your own? Now think about how your story characters would use multiplication, division, and fractions. What sorts of things would they compare, or arrange into groups, or share between friends?

[11]

Write a scaling-up problem set in the world of a book or movie you enjoy. There will be one group or part or thing that is some number of times as many or as big as something else.

Your problem has four numbers:

+ The smaller amount.

+ The larger amount.

+ The scale factor (how many times as much).

+ And the total of both amounts.

Tell any two of these numbers. Then ask your reader to figure out one (or both) of the others.

[12]

Write a multistep comparison problem set in the world of a book or movie you enjoy. Your story needs four numbers:

+ A smaller amount.

+ A larger amount.

+ The difference between them.

+ The total of both amounts.

Tell the difference and the total amount. Can your reader figure out the size of the items?

[13]

Write a problem about a collection, set, or group composed of three or four parts.

The numbers may include the total amount and the sizes of the parts. Or instead of telling the size of a part directly, you might use a comparison (how much more or less, or how many times as much) to make a multistep puzzle-within-a-problem.

How much information do you need to include so that your reader can figure it out?

[14]

Write a problem that contains some sort of this-per-that unit. For example, where in your story world would people think about the number of cups per pint, candies per package, cost per drink, or something like that?

Your problem has three numbers:

♦ How many same-size units.

♦ The size of a single unit.

♦ The total amount.

Tell any two of these numbers. Then ask your reader to figure out the third.

[15]

Write a problem about sharing something by dividing it. Use fractions in the problem, or let the answer be a fraction.

[16]

Write a problem with a scaling-down comparison. Make one group or item some fraction of the other.

[17]

Write a problem that includes some type of measurement with a mixed number.

[18]

Write a problem that uses fractions with different denominators.

For an easier problem, make the denominators related to each other, like thirds and sixths. For a harder problem, use unrelated denominators.

[19]

Write a story using fraction division. Tell the size of a fractional part of your quantity or group. Then ask readers to find the whole amount.

[20]

Write any kind of problem you like.

Extend Your Skills: Measurement and Decimals

5

Word Problems Inspired by *The Railway Children*

A mother and three children move to the countryside and meet new neighbors.

MANY CHILDREN ARE CONFUSED BY decimals. They are convinced 0.48 is greater than 0.6 because forty-eight is obviously ever so much bigger than six. Their intuition tells them that "point-2 times point-3 is point-6" has the clear ring of truth. And they confidently assert that, if you want to multiply a decimal number by ten, all you need to do is add a zero at the end.

What can we do to help our kids understand decimals? Here are several word problem scenarios to try, set back in the time of Edith Nesbit's *The Railway Children*.

See how many bar model diagrams you can draw before reading my solutions.

[1]

When Peter was sick, Mother gave him barley water to drink. Phyllis was curious and wanted a taste also. Peter drank 0.7 liter of barley water. Phyllis drank 0.2 liter.

(a) How much barley water did they drink?
(b) How much more did Peter drink than Phyllis?

[2]

The children made a bandit's lair in the attic. Peter and Bobbie captured Phyllis to hold her for ransom. Phyllis insisted on being tied up. Phyllis had a piece of string 1.85 m long. Then Bobbie found another piece of string that was 1.4 m longer than the first. Find the total length of the two pieces of string.

[3]

When mother was ill, the old gentleman bought a good many gifts and packed them in a hamper to send to the children. At one shop, he bought 1 kg of grapes for £4.15 and a basket of peaches for £7.59. He gave the cashier £15. How much change did the old gentleman receive?

[4]

On Bobbie's birthday, the family had fancy cranberry juice along with their tea. Bobbie, Peter, and Phyllis each drank 0.4 liter of juice. How many liters of juice did the children drink in all?

[5]

Mother had £30. She wanted to take her children to a nearby town for an outing. She bought 4 train tickets. Each ticket cost £2.95. How much money did Mother have left?

[6]

The Doctor had 0.9 liter of medicine at his house. He poured it into 3 bottles and gave one to the sick Russian stranger. How much medicine was in the little bottle?

[7]

After the children saved the train, Mother bought flannel to make warm, new petticoats. She bought 5 meters of flannel for £8. How much did one meter of flannel cost?

[8]

The Station Master kept boxes of fancy chocolate bars that passengers could buy to eat while they waited for their train. A full box of 12 bars weighed 2.34 kg.

After the ceremony with the watches, the Station Master gave each of the children 4 whole bars of chocolate. Then the empty box weighed 0.06 kg. Find the weight of one chocolate bar.

[9]

Mrs. Ransome posted 4 parcels. One of them weighed 1.8 kg. The other 3 parcels weighed 2.05 kg each. Find the total weight of the parcels.

[10]

Peter helped the grocer carry boxes into his storeroom. One box contained 5 packets of flour. If the total weight of the flour was 5 kg 650 g, find the weight of each packet of flour.

[11]

Where the landslide covered the railway, it took the workmen 7 h 30 min to clear away the debris from 6 meters of the railway track. How long did it take for the workmen to clear one meter?

[12]

Bobbie and Jim told each other stories to take their minds off the darkness in the tunnel. Jim told how he'd once helped to build a stone fence. He spent 3 h 30 min every morning, carrying stones and lifting them into place. He finished the fence in 5 mornings. How much time did Jim spend working on the fence?

[13]

After the old gentleman turned their house into the Three Chimneys Hospital, Mother sewed new curtains for the parlor and the 3 bedrooms. She used 3.46 m of material for each bedroom and 4.25 m of material for the parlor. How much material did mother use altogether?

[14]

The usual price of a large cinnamon sticky bun was £0.60. The baker had a sale, and mother bought 4 sticky buns for Jim and the children to have with tea. Mother paid £2.20 for the buns. How much cheaper was each sticky bun during the sale?

Bar Models and Answers

Key Tips for Decimals

The word *decimal* means "base ten." The place value relationships of our number system are based on the number 10, as follows:

♦ 10 ones/units = 1 ten; 1 unit = $\frac{1}{10}$ ten

♦ 10 tens = 1 hundred; 1 ten = $\frac{1}{10}$ hundred

♦ 10 hundreds = 1 thousand; 1 hundred = $\frac{1}{10}$ thousand

♦ and so on

In each instance, a digit in one place value column is worth ten times the value of the same digit in the column to its right and $\frac{1}{10}$ the value of the same digit in the column to its left. So by definition, all our numbers are decimal numbers.

But most people use the word *decimal* as a shortcut for *decimal fractions*, the numbers that come after the decimal point. The place value pattern remains the same for decimal fractions.

♦ 10 tenths = 1 one/unit; 1 tenth = $\frac{1}{10}$ unit

♦ 10 hundredths = 1 tenth; 1 hundredth = $\frac{1}{10}$ tenth

♦ 10 thousandths = 1 hundredth; 1 thousandth = $\frac{1}{10}$ hundredth

♦ and so on

The numbers to the right of the decimal point don't *look* like fractions, because their denominators are invisible. But they *are* fractions, which means they will often act in ways children don't expect, especially when using multiplication or division. You can prevent many common mistakes by reminding your students

◁ Wholenumberville . Fractionland ▷

that the decimal numbers are fractions.

When thinking about fractions, our children lean heavily on their real-life experiences of sharing toys or food. But most of us divide things into simple fractions, cutting a sandwich into halves, splitting the last piece of cake three ways, sharing a sheet of stickers among a small group of friends.

Elementary children have almost no experience of tenths or hundredths. Even with money, most children don't consider dimes or pennies as fractions of a dollar. They normally think of pennies as the primary counting unit, with dollars being a collection of cents. Or they think of centimeters and meters as being different primary units with an easy conversion between them.

To get our children truly thinking about decimal fractions, we push them to reason more deeply about metric measurement and about money.

Teach children these tips for working with decimals:

1) Remember the invisible denominators of your decimal fractions.

2) Break apart and combine numbers to make your calculations easier.

3) Use your common sense.

Make sure your children gain a solid understanding of ordinary fraction calculations before expecting them to work with decimals.

And don't teach the standard *algorithms* (procedural rules) until the children have plenty of practice making sense of decimal fractions on their own. Most students won't need the algorithms at all, since in real life we use a calculator for any calculations too complex to work mentally. Algorithms are easy to misremember, but if students build a solid foundation of sense-making, they can fall back on that whenever they forget a rule.

[1]

> *When Peter was sick, Mother gave him barley water to drink.*
> *Phyllis was curious and wanted a taste also. Peter drank 0.7*
> *liter of barley water. Phyllis drank 0.2 liter.*
> *(a) How much barley water did they drink?*
> *(b) How much more did Peter drink than Phyllis?*

We begin with simple tenths, and we limit ourselves to addition or subtraction, so at first decimals seem to act the same as whole numbers, making intuition a reliable guide. Here we add tenths to tenths to find an answer that's also in tenths.

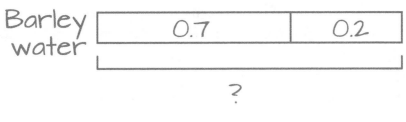

$$0.7 + 0.2 = 7 \text{ tenths} + 2 \text{ tenths}$$
$$= 9 \text{ tenths} = 0.9$$

Together, the children drank 0.9 liter of barley water.

This problem had two questions. In subtraction, we find the difference between our two parts. Notice the algebraic power of the bar model: The blocks are variables. Even when they're not drawn to scale, the model works as a thinking tool.

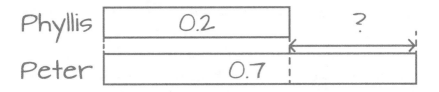

$$0.7 - 0.2 = 7 \text{ tenths} - 2 \text{ tenths}$$
$$= 5 \text{ tenths} = 0.5$$

Peter drank 0.5 liter more than Phyllis drank.

[2]

The children made a bandit's lair in the attic. Peter and Bobbie captured Phyllis to hold her for ransom. Phyllis insisted on being tied up. Phyllis had a piece of string 1.85 m long. Then Bobbie found another piece of string that was 1.4 m longer than the first. Find the total length of the two pieces of string.

Now we move farther into Fractionland, dealing with hundredths of a meter. Still, children can lean on their understanding of whole-number base ten relationships to make sense of decimal fractions. Just as the number 300 may be thought of as 3 hundreds or 30 tens or 300 ones, so we can look at decimals in the same flexible way, choosing the relationship that best helps us make sense of our problem:

$$1.85 \text{ m} = 1 \text{ m} + 85 \text{ hundredths}$$
$$= 1 \text{ m} + 8 \text{ tenths} + 5 \text{ hundredths}$$
$$= 18 \text{ tenths} + 5 \text{ hundredths}$$
$$= 185 \text{ hundredths of a meter}$$

But now we discover the first way decimal fractions confuse students.

With whole number place value, students learned the importance of adding like to like: ones to ones, tens to tens, hundreds to hundreds, and so on. They simply lined up the ones (or units) column on the right, and all the other values fell into line.

But with decimal fractions, the ones column can be anywhere in the middle of our number, and the digits on the right may be different types: tenths, hundredths, thousandths, etc. Yet we still need to add like to like. Our children must look for the decimal point as a reliable guide to matching the place values so they can add tenths to tenths, hundredths to hundredths, and so on.

Also, this problem presents a challenge unrelated to decimal fractions: A careless student will read the two numbers as the lengths of their respective strings, but the second number was only part of the length.

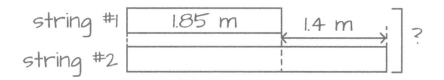

$$1.85 = \text{one meter, 8 tenths, and 5 hundredths}$$
$$\text{and}$$
$$1.4 = \text{one meter and 4 tenths}$$
$$\text{so}$$
$$1.85 + 1.4 = 2 \text{ whole meters,}$$
$$\text{plus } (8 + 4 = 12) \text{ tenths,}$$
$$\text{plus 5 hundredths}$$

And since 10 tenths makes another whole meter, the second piece of string was 3.25 m long.

Finally, we add the two strings together:

$$1.85 + 3.25 = 4 \text{ meters} + 10 \text{ tenths} + 10 \text{ hundredths} = 5.1$$

The total length of the two pieces of string was 5.1 m.

[3]

When mother was ill, the old gentleman bought a good many gifts and packed them in a hamper to send to the children. At one shop, he bought 1 kg of grapes for £4.15 and a basket of peaches for £7.59. He gave the cashier £15. How much change did the old gentleman receive?

Counting change provides a natural context for decimal subtraction puzzles. But children often treat pennies and dollars (or pence and pounds) as separate units of measurement, not as a whole unit plus fractional parts.

We must draw their attention to the decimal fraction relationships and encourage them to use their understanding of money in making sense of decimals. In this problem, for example, they know that £15 may be written as £15.00. They must rename one of these pounds as 100 pence before they can do the subtraction.

Also, as students mature in their problem-solving skills, their textbooks offer them additional challenges. For example, situations in the real world present an abundance of information from which we must sort out the data we need.

So here, the problem offers a bit of unnecessary information. A careless student will simply grab all the numbers and start calculating. But a student who applies common sense will realize the weight of the grapes is irrelevant to the old gentleman's bill.

bill	7.59	4.15	?
cash	15		

$$4.15 + 7.59$$
$$= 11 \text{ and } (1 + 5) \text{ tenths and}$$
$$(5 + 9) \text{ hundredths}$$
$$= 11 \text{ and } 6 \text{ tenths and}$$
$$14 \text{ hundredths}$$
$$= 11.74$$
… and then:
$$15.00 - 11.74$$
$$= (14 + 1.00) - 11.74$$
$$= (14 - 11) \text{ and } (100 - 74)$$
$$\text{hundredths}$$
$$= 3.26$$

The old gentleman received £3.26 in change.

[4]

On Bobbie's birthday, the family had fancy cranberry juice along with their tea. Bobbie, Peter, and Phyllis each drank 0.4 liter of juice. How many liters of juice did the children drink in all?

When we move to a new topic, we often go back to simpler numbers. This helps students concentrate on making sense of the new idea without being distracted by the thorns and thistles of arithmetic. So here, as we transition to multiplication, we go back to using only tenths.

Also, we don't jump directly to a decimal-times-decimal problem. Math monsters like "0.2 × 0.3" are still far in the future.

Yet this relatively basic problem is enough to confuse many students, who will want to believe that three times point-4 should make point-twelve. As a reality check, estimate the answer with fractions before calculating with decimals. Four-tenths is almost one-half, so the final answer should be a bit less than three-halves, which is 1.5.

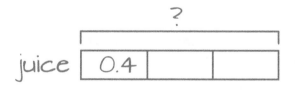

One unit = 0.4
3 units = 3 × 0.4 = 3 × 4 tenths
= 12 tenths = 1.2

The children drank 1.2 liters of juice in all.

NOTE ON THE MATH MONSTER: What will you do when your children eventually need to deal with monster calculations like

"0.2 × 0.3"? Even then, thinking in fractions remains more helpful to your students than trying to teach them easily misremembered rules about moving the decimal point. How much is ²⁄₁₀ of ³⁄₁₀? Can they see that it must be a tiny fraction? Certainly not 0.6!

You can use a *ratio table* (a chart of equivalent fractions; see Chapter 7 for more details) to calculate decimal multiplication by scaling. Start with one of whatever you are multiplying, in this case 0.3. Then scale the ratio to get the amount you want, in this case 0.2 of it.

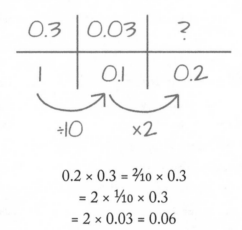

$$0.2 \times 0.3 = \tfrac{2}{10} \times 0.3$$
$$= 2 \times \tfrac{1}{10} \times 0.3$$
$$= 2 \times 0.03 = 0.06$$

[5]

Mother had £30. She wanted to take her children to a nearby town for an outing. She bought 4 train tickets. Each ticket cost £2.95. How much money did Mother have left?

When we're working with fractions, multiplication can do funny things. Encourage your students to develop the habit of estimating before they calculate, using common sense to protect them from place-value errors.

$$\text{One unit} = 2.95$$
$$\text{4 units} = 4 \times 2.95$$
$$= 4 \times \text{almost 3} = \text{almost 12}$$

tickets	2.95			?
money	30			

Then do your exact calculations:

$$4 \times 2.95 = 4 \times 2 \text{ and } 4 \times 95 \text{ hundredths}$$
$$= 8 \text{ and } 380 \text{ hundredths} = 11.80$$
$$(\text{Check: } 11.80 = \text{almost } 12? \text{ Yes.})$$
$$\dots \text{ and then:}$$
$$30 - 11.80 = \text{a bit more than } (30 - 12 = 18)$$
$$30 - 11.80 = 18.20$$

Mother had £18.20 left after buying the tickets.

[6]

The Doctor had 0.9 liter of medicine at his house. He poured it into 3 bottles and gave one to the sick Russian stranger. How much medicine was in the little bottle?

Division is the inverse of multiplication: Instead of grouping together multiple units, we share out same-size portions.

Because division is a new topic, we return to easy numbers. Students understand the relationship 9 ÷ 3 and can use this in thinking about decimal fractions. Nine of anything, shared

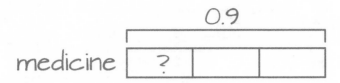

between three people, will give each person three of that thing. It doesn't matter whether we are sharing out candy or books or tenths of a liter of medicine.

$$3 \text{ units} = 0.9$$
$$\text{One unit} = 0.9 \div 3 = 9 \text{ tenths} \div 3$$
$$= 3 \text{ tenths} = 0.3$$

Each little bottle held 0.3 liter of medicine.

[7]

After the children saved the train, Mother bought flannel to make warm, new petticoats. She bought 5 meters of flannel for £8. How much did one meter of flannel cost?

Children could use long division to solve this problem, but working mentally is a better way to strengthen their mastery of decimal fractions. As much as possible, encourage your children to do their math mentally. Not by trying to replicate pencil-and-paper steps in their minds, but by thinking about the numbers and how they behave.

To divide £8 mentally, we can imagine splitting the money into five piles:

8

flannel [? | | | | |]

- ◆ First put one pound in each pile. Now we have £3 left.

- ◆ Put one-half pound in each pile, which leaves £0.50 to split.

- ◆ Put ten pence more per pile.

$$\text{One unit} = 8 \div 5 = 1 + 0.50 + 0.10 = 1.60$$

Each meter of flannel cost £1.60.

[8]

The Station Master kept boxes of fancy chocolate bars that passengers could buy to eat while they waited for their train. A full box of 12 bars weighed 2.34 kg. After the ceremony with the watches, the Station Master gave each of the children 4 whole bars of chocolate. Then the empty box weighed 0.06 kg. Find the weight of one chocolate bar.

You may want to remind students that 3 tenths and 4 hundredths is the same as 34 hundredths. This makes the subtraction easier to do by mental math, since we can think "34 – 6."

empty box [0.06]
chocolate [?]
] 2.34

$$2.34 - 0.06 = (2 \text{ and } 34 \text{ hundredths}) - 6 \text{ hundredths}$$
$$= 2 \text{ and } 28 \text{ hundredths} = 2.28$$

The total weight of the chocolate is 2.28 kg. That's for 12 bars, so we need division to give us the weight of a single bar.

As the numbers get more complicated, long division may

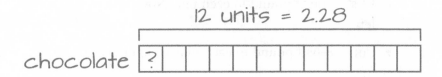

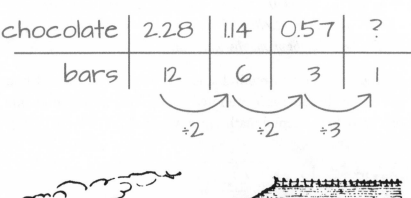

be a useful tool for some students. If you struggle to explain long division in a way that makes sense (not just "follow these steps"), read my blog post "The Cookie Factory Guide to Long Division."[†]

Other students will prefer to do most of the work mentally, using scratch paper to keep track of their progress. A ratio table can help. Start with the ratio of 2.28 kg to 12 units, and then scale the ratio to find the value of one unit. (For a more thorough explanation of ratio tables, see Chapter 7.)

chocolate	2.28	1.14	0.57	?
bars	12	6	3	1

÷2   ÷2   ÷3

† *denisegaskins.com/the-cookie-factory-guide-to-long-division*

$$12 \text{ units} = 2.28$$
$$\text{One unit} = 2.28 \div 12$$
$$(\text{estimate: less than } 3 \div 12 = \tfrac{1}{4} = 0.25)$$
$$= 2.28 \div (2 \times 2 \times 3)$$
$$= ((2.28 \div 2) \div 2) \div 3$$
$$= (1.14 \div 2) \div 3$$
$$= 0.57 \div 3$$
$$= (0.60 - 0.03) \div 3$$
$$= (0.60 \div 3) - (0.03 \div 3)$$
$$= 0.20 - .0.01 = 0.19$$

One chocolate bar weighed 0.19 kg.

LONG DIVISION TIP: The long division algorithm *only* works when you are *dividing by a whole number.* Therefore, if you want to divide by a decimal fraction, you need to convert your calculation into an equivalent division by a whole number.

The initial "move the decimal point" step of long division is like making an equivalent fraction:

$$3.59 \div 1.7 = ?$$
$$\tfrac{3.59}{1.7} \times \tfrac{10}{10} = \tfrac{35.9}{17}$$
$$= 35.9 \div 17$$

Decimal points *never* move! They always sit directly between Wholenumberville and Fractionland. When you hear a procedural rule say "move the decimal point," always look closer at the calculation and try to understand what's really happening.

[9]

Mrs. Ransome posted 4 parcels. One of them weighed 1.8 kg. The other 3 parcels weighed 2.05 kg each. Find the total weight of the parcels.

Have you noticed that the structure of all these problems is much simpler than, for example, the wolves-and-dwarfs problem in Chapter 4? It doesn't take much thought to diagram these relationships.

At this level, we can challenge students with the logic of a puzzle or with the difficulty of a calculation. But we try not to do both at once, to avoid frustration and discouragement.

Decimal fractions are a new topic that may confuse students, so we keep the structure of the problems relatively easy to follow—basic addition or subtraction, simple groups for multiplication or sharing for division.

one parcel $\boxed{1.8}$ $\Big]$?

3 parcels $\boxed{2.05}\ \boxed{}\ \boxed{}$ $\Big]$

One unit = 2.05
3 units = 3 × 2.05 = 3 × (2 and 5 hundredths)
= (3 × 2) and (3 × 5) hundredths
= 6 and 15 hundredths = 6.15
… and then:
Total = 6.15 + 1.8
= (6 + 1) kg and (1 + 8) tenths and 5 hundredths
= 7.95

The total weight of all four parcels was 7.95 kg.

[10]

Peter helped the grocer carry boxes into his storeroom. One box contained 5 packets of flour. If the total weight of the flour was 5 kg 650 g, find the weight of each packet of flour.

Measurements offer a natural way to start building intuition for more challenging decimal calculations. Students could deal

5.650

flour | ? | | | | |

with the kilograms and grams independently, but ask them to think of the whole amount as "5.650 kg."

To simplify it further, point out that students can always ignore a zero at the end of a decimal fraction. There are 0 thousandths, just as there are 0 ten-thousandths and 0 millionths and zero of many other tiny fractions.

$$5 \text{ units} = 5.650 = 5.65$$
$$\text{One unit} = 5.65 \div 5 = 1.13$$

Each packet of flour weighed 1.13 kg, or 1 kg 130 g.

[11]

Where the landslide covered the railway, it took the workmen 7 h 30 min to clear away the debris from 6 meters of the railway track. How long did it take for the workmen to clear one meter?

7 hr 30 min

track (m) | ? | | | | | |

This problem trips up many students because time is *not* a decimal number. We measure things in many different ways—as students will discover to their chagrin in high school chemistry—and converting from one measurement unit to another requires thinking about their relationships. Students who try to work by rote, without thinking, will often convert 7 h 30 min to 7.3 hours.

$$6 \text{ m} = \text{cleared in 7 h 30 min}$$
$$\text{One meter} = \text{cleared in 7 h 30 min} \div 6$$
$$= \text{cleared in } (6 + 1) \text{ h 30 min} \div 6$$
$$= (6 \text{ h} \div 6) \text{ and } (1 \text{ h 30 min} \div 6)$$
$$= 1 \text{ h and } (90 \text{ min} \div 6)$$
$$= 1 \text{ h 15 min}$$

It took 1 h 15 min for the workmen to clear one meter of track.

[12]

Bobbie and Jim told each other stories to take their minds off the darkness in the tunnel. Jim told how he'd once helped to build a stone fence. He spent 3 h 30 min every morning, carrying stones and lifting them into place. He finished the fence in 5 mornings. How much time did Jim spend working on the fence?

?

3h 30min				

This problem offers a second chance to students who got confused by the previous one.

$$\text{One unit} = 3 \text{ h 30 min}$$
$$5 \text{ units} = 3 \text{ h 30 min} \times 5$$
$$= 15 \text{ h 150 min}$$
$$= 15 \text{ h and } (60 + 60 + 30) \text{ min}$$
$$= 17 \text{ h 30 min}$$

Jim spent 17 h 30 min working on the fence.

[13]

After the old gentleman turned their house into the Three Chimneys Hospital, Mother sewed new curtains for the parlor and the 3 bedrooms. She used 3.46 m of material for each bedroom and 4.25 m of material for the parlor. How much material did mother use altogether?

parlor | 4.25 |
bedrooms | 3.46 | | | ?

One unit = 3.46
3 units = 3 × 3.46
= (3 × 3) + (3 × 46 hundredths)
= 9 + 138 hundredths
= 9 + 1 + 38 hundredths = 10.38
… and then:
Total = 10.38 + 4.25 = 14.63

Mother used 14.63 m of material altogether.

[14]

The usual price of a large cinnamon sticky bun was £0.60. The baker had a sale, and mother bought 4 sticky buns for Jim and the children to have with tea. Mother paid £2.20 for the buns. How much cheaper was each sticky bun during the sale?

There are two ways to approach this problem. (There are probably more ways, since math is like that.) I might first calculate the new price and then see how much cheaper it is, as follows.

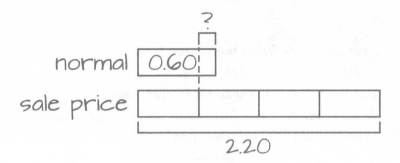

4 sale-price buns = 2.20
One sale-price bun = 2.20 ÷ 4 = 0.55
Discount per bun = 0.60 – 0.55 = 0.05

On the other hand, my daughter might calculate the discount directly, without bothering to find the individual price.

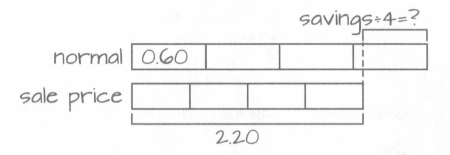

One list-price bun = 0.60
4 list-price buns = 4 × 0.60 = 2.40
Savings = 2.40 – 2.20 = 0.20
Discount per bun = 0.20 ÷ 4 = 0.05

When we find independent ways to solve a problem—which happens far more often than I would have expected—and the answers don't match, we know at least one of us has made a mistake. But if we confirm each other's answer, we can be pretty sure we're right even without checking the back of the book.

Whatever way you solve it, Mother saved 5 pence on each sticky bun.

Make Your Own Math

Where in your favorite story world would the characters use decimal fractions? Do they measure distances or weights, or do they have a decimal-based money system? What kind of math stories will you create?

[15]

Write an addition story with decimal measurements.

Instead of telling the size of a part directly, you might use a comparison (how much more or less, or how many times as much) to make a multistep puzzle-within-a-problem.

How much information do you need to include so that your reader can figure it out?

[16]

Write a subtraction problem with decimals, perhaps by counting money and making change. The problem may also contain addition, as a normal shopping trip might.

[17]

Write a multistep comparison problem using decimals. Your story has four (or five) numbers:

♦ A smaller amount.

♦ A larger amount.

♦ The difference between them.

♦ (Optional) How many times the larger number is of the smaller number's size, or what fraction the smaller number is of the larger.

♦ The total of both amounts.

Tell any two of the numbers, and then ask for whichever of the missing quantities you like.

[18]

Write a problem that involves multiplication with three or more decimal-fraction-sized parts. You can ask for the total product or ask for a sum or difference involving that product.

[19]

Write a problem about sharing something by dividing it. Use decimals in the problem, or let the answer be a decimal.

[20]

Write a problem involving time that requires the reader to convert between minutes and hours, or between seconds and minutes. Can you trick them into using a decimal conversion instead of the correct base-60?

[21]

Write a problem about dividing something by measuring out mixed-decimal-fraction parts. That is, measure out parts that are greater than one but have a decimal fraction at the end (for example, 3.75). Your story has four numbers:

- ◆ The original amount (which may be a whole number or have a decimal part).
- ◆ The size of the measured-out parts (a decimal mixed number).
- ◆ The number of parts you get (a whole number).
- ◆ The leftover amount, if any.

Tell any two of the numbers, and then ask for one (or both) of the missing quantities.

[22]

Write any kind of problem you like.

Reap the Reward:
Ratios and
More Fractions

Word Problems Inspired by *The Hobbit*

*A halfling joins dwarfs and a wizard on the
quest to steal a dragon's treasure.*

As UPPER-ELEMENTARY STUDENTS APPROACH MIDDLE school,
they have learned the foundations of arithmetic: addition, subtraction, multiplication, division, and basic fractions. Now they
need to master fraction multiplication and play around with
ratios—two concepts that will eventually merge to become
algebraic proportions. Word problems grow ever more complex,
and learning to explain multistep solutions becomes a first step
toward writing proofs.

For your reading pleasure, I've created stories that might happen in a medieval fantasy world like that of J. R. R. Tolkien's
classic, *The Hobbit.*

See how many bar model diagrams you can draw before reading my solutions.

[1]

> *The halfling had 3 times as many apple tarts as mince pies
> in his larder. If he had 24 more apple tarts than mince pies,*

how many of the pastries (both tarts and pies) did he have altogether?

[2]

Three trolls had 123 pieces of gold. Andrew had 15 pieces of gold more than Bert. Bert had 3 pieces fewer than Charles. How many pieces of gold did Charles have?

[3]

The Great Goblin had twice as many goblin soldiers as his cousin, the Gross Goblin. How many soldiers must the Great Goblin send to his cousin so that they will each have 1,200 goblin soldiers?

[4]

The cave creature caught 10 small fish. He divided the fish to make 4 equal meals. How many fish did he eat at each meal?

[5]

The giant bear-man baked a large loaf of whole-grain bread. He ate ⅓ of the loaf himself (with plenty of honey), and he sliced ½ of the same loaf to feed the dwarfs and their halfling friend. What fraction of the loaf was left?

[6]

The king of the elves had a barrel of fine wine. His butler poured ¾ gallon of it into a small keg. He drank ½ of the keg and gave the other half to his friend, the chief of the guards. How much wine did the elven king's butler drink?

[7]

⅔ of the items in the dragon's treasure were made of gold. ¼ of the remaining part was precious gems. What fraction of the dragon's treasure was precious gems?

[8]

When the king of the elves heard the dragon had been killed, he set out to claim a share of the treasure. ⅖ of his army were archers. ½ of the remainder fought with spears, and the rest carried swords. If 300 soldiers carried swords, how many elves marched out with the elven king?

[9]

The master bowman gathered a small army of 600 survivors from the town the dragon had destroyed. The ratio of archers to swordsmen was 2:3. How many archers followed the master bowman to the final battle?

[10]

The dwarfs rewarded their halfling friend with two chests of gold, silver, and small gems—6,000 pieces of treasure altogether. There were twice as many pieces of gold as there were gems. There were 600 more pieces of silver than gems. How much of each type of treasure did the halfling receive?

Bar Models and Answers

[1]

> *The halfling had 3 times as many apple tarts as mince pies in his larder. If he had 24 more apple tarts than mince pies, how many of the pastries (both tarts and pies) did he have altogether?*

Long before the textbook explicitly introduces ratios and proportional reasoning, students get plenty of practice with informal ratios like *twice as many* and *three times as many*. In each case, we draw the smaller group as one unit, and the larger group as whatever number of units make the "times as many."

In this problem, the unit will be the number of mince pies. Our halfling has twenty-four more tarts than pies, and we are looking for the total.

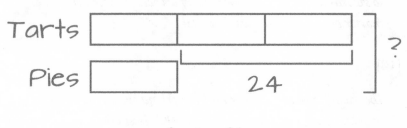

$$2 \text{ units} = 24$$
$$1 \text{ unit} = 24 \div 2 = 12$$
$$4 \text{ units} = 4 \times 12 = 48$$

There are forty-eight pastries altogether.

[2]

> *Three trolls had 123 pieces of gold. Andrew had 15 pieces of gold more than Bert. Bert had 3 pieces fewer than Charles. How many pieces of gold did Charles have?*

This is a comparison problem. We will need three bars—one for each troll's bag of gold. Since Bert is compared to each of the others, we'll put his bar in the middle.

Notice that Bert has the smallest number of gold pieces. Andrew has Bert's amount plus fifteen more, and Charles has Bert's amount plus three more. And the problem tells us the total number of gold pieces.

Andrew [15]
Bert   [   ] ⎤ 123
Charles [3]⎦

We could make an algebra equation: $3x + 18 = 123$. But instead, we'll use the "stealth algebra" of our diagram to help us think through the number relationships.

Together, the trolls have 123 pieces of gold. We can imagine taking away the extra coins from Andrew's and Charles's bars, reducing each of their loot to match Bert's stash.

$$3 \text{ units} = 123 - 18 = 105$$
$$1 \text{ unit} = 105 \div 3 = 35$$

And now, because I forgot to put a question mark in my drawing, I have to read the problem again. Oh, yes: We need to find Charles's amount of gold.

$$1 \text{ unit} + 3 = 35 + 3 = 38$$

Charles had thirty-eight gold pieces.

[3]

The Great Goblin had twice as many goblin soldiers as his cousin, the Gross Goblin. How many soldiers must the Great Goblin send to his cousin so that they will each have 1,200 goblin soldiers?

What a cool problem! It seems like they haven't given enough information, doesn't it? Students who are not used to bar diagrams often get confused by transfer problems, which have a beginning situation and then something gets moved from one person to another to set up the end of the story. Desperate children will grab any numbers and guess at the answer.

Let's see what the bar diagrams tell us. First, the starting ratio: One goblin has twice as many soldiers as the other.

Great Goblin [|]
Gross Goblin [   ]

And below is our final situation: Both armies have 1,200 goblin soldiers.

Great Goblin [1200   ]
Gross Goblin [1200   ]

But how can we get from the initial situation to our goal?

A useful problem-solving tool is to work backwards. Here we'll start with the end of the story, since that's where they gave us a number to work with. We know how many goblins will be in each army, which means we can find the total number of soldiers.

$$1200 \times 2 = 2400$$

There are 2,400 goblin soldiers in the final diagram. And moving the soldiers around won't change the total number, so there must have been 2,400 soldiers in the first diagram, too.

Great Goblin

Gross Goblin

2400

$$3 \text{ units} = 2400$$
$$1 \text{ unit} = 2400 \div 3 = 800$$
$$2 \text{ units} = 2 \times 800 = 1600$$

So at the beginning, the Great Goblin has 1,600 soldiers, while his cousin has 800. To make them balance, the Great Gob must cut his army down to 1,200.

$$1600 - 1200 = 400$$

The Great Goblin sent four hundred soldiers to his cousin.

Alternate Solution

There are many ways to approach a word problem. My daughter used a shorter method than the way I did it. She started with the simple 2:1 ratio at the beginning of the story.

But how many soldiers need to move? She could see that the Great Goblin has to send one-fourth of his soldiers to make the two armies equal. We need to move half a unit, so that the soldiers who move match the other half who stay.

Great Goblin
Gross Goblin

Since the "extra" unit had to divide in half, she divided all the other units in half, too. Then she realized that the three smaller units remaining in the Great Goblin's army must be the 1,200 soldiers he has at the end of the story.

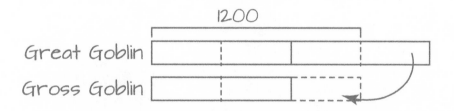

1200

Great Goblin
Gross Goblin

3 small units = 1200
1 small unit = 1200 ÷ 3 = 400

While the process may vary, the answer must be the same. The Great Goblin sent four hundred soldiers to his cousin.

[4]

> *The cave creature caught 10 small fish. He divided the fish to make 4 equal meals. How many fish did he eat at each meal?*

Word problems like this one make explicit something that in earlier grades has only been implied—the link between fractions and division. I tell my students, "The division symbol looks like a tiny fraction, with dots for the numbers. Let that remind you: Every fraction is a division problem, top divided by bottom. And we can write any division problem as a fraction."

Our cave creature has a row of fish, split into four equal groups. Each group is one-fourth of the whole amount.

?

Fish

10

$$1 \text{ unit} = 10 \div 4 = {}^{10}\!/_4$$
$${}^{10}\!/_4 = {}^5\!/_2 = 2\tfrac{1}{2}$$

The cave creature ate two and one-half fish at each meal.

[5]

> *The giant bear-man baked a large loaf of whole-grain bread. He ate ⅓ of the loaf himself (with plenty of honey), and he sliced ½ of the same loaf to feed the dwarfs and their halfling friend. What fraction of the loaf was left?*

Bar diagrams help students see the need for a common denominator. Unless the pieces are the same size, it doesn't make much sense to say, "We have one piece left."

The bar is the bear-man's loaf of bread. Imagine that he cut one-third of the loaf from one side, and half of the loaf from the other. Then the piece that's left in the middle is our mystery fraction.

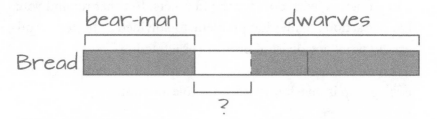

What size is the white piece in the middle? It is clearly half of the middle third because the portions on each side of the dashed "half-loaf" line must be equal. Let's get a common denominator by cutting all the thirds in half.

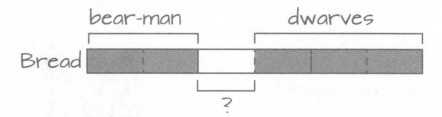

Aha! Half of a third is the same as a sixth.

$$1 - \tfrac{1}{3} - \tfrac{1}{2} = 1 - \tfrac{5}{6} = \tfrac{1}{6}$$

The bear-man had one-sixth of his loaf of bread left.

[6]

The king of the elves had a barrel of fine wine. His butler poured ¾ gallon of it into a small keg. He drank ½ of the keg

and gave the other half to his friend, the chief of the guards. How much wine did the elven king's butler drink?

I teach my students to think of multiplication when they hear the preposition "of." In this problem, we want half *of* three-fourths gallon, which is the following calculation.

$$\tfrac{1}{2} \times \tfrac{3}{4} = ?$$

This could be drawn on a single bar, but I find it easier to draw two. The first bar represents a full gallon of wine. The second bar is the small keg from which the butler and his friend are sharing.

gallon

Wine				

Keg	?	

Notice that the line that splits the keg in half also bisects the unit above it, the second fourth of the gallon. Let's make an equivalent fraction by splitting all the units of wine in half:

gallon

Wine								

Keg	?	

Now we can see that the keg holds six-eighths of a gallon of wine. Each of the elves drank half the keg.

½ of ¾ = ½ of ⁶⁄₈ = ⅜

The butler drank three-eighths of a gallon of wine.

[7]

> ⅔ *of the items in the dragon's treasure were made of gold. ¼ of the remaining part was precious gems. What fraction of the dragon's treasure was precious gems?*

As students gain skill in working with fractions, the problems grow more complex. Now we introduce the "fraction of the remaining part" problem. In this type of problem, what began as a part of the original whole becomes a new "whole thing" we can cut into parts of its own.

This foreshadows the work we'll do later with percents, where the *base* (what we count as 100%) is often something less than "everything there is."

The dragon's treasure is divided into thirds, and one of these thirds is our remaining part. We show that it is being treated separately by drawing a new bar below the first, connected with lines to its original position.

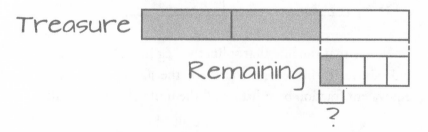

We are interested in one-fourth of the remaining part, and we want to know what fraction of the original bar (the whole treasure) it would be.

Well, since we cut our remaining piece into fourths, we could make a common denominator by cutting all the thirds the same way. What size pieces would we have then?

There would be 3 × 4 = 12 of them in the whole bar.

$$¼ \text{ of } ⅓ = ¼ × ⅓$$
$$= ¼ × ⁴/₁₂ = ¹/₁₂$$

One-twelfth of the dragon's treasure was precious gems.

[8]

When the king of the elves heard the dragon had been killed, he set out to claim a share of the treasure. ⅖ of his army were archers. ½ of the remainder fought with spears, and the rest carried swords. If 300 soldiers carried swords, how many elves marched out with the elven king?

Here is another fraction-of-the-remainder puzzle, but this time the denominators are not related. We begin with a bar for the elven king's army divided into fifths, two of which are archers. The remaining three-fifths becomes a new whole bar, of which half fight with spears.

?

Elves

Remaining   300

We could simplify to a single unit (common denominator) by dividing the original bar into tenths and the remainder bar into sixths. But small pieces like tenths are hard to handle. It's easier to work with each bar separately.

Since the remainder bar has a number, that's where we'll start. There are three hundred swordsmen. (Swords-elves?) That means we can find the size of our remaining part.

$$1 \text{ unit} = 300$$
$$2 \text{ units} = 300 \times 2 = 600$$

If the remaining part of the army is six hundred, then on the original bar:

$$3 \text{ small units} = 600$$
$$1 \text{ small unit} = 600 \div 3 = 200$$
$$5 \text{ small units} = 200 \times 5 = 1000$$

There were a thousand elves marching with their king.

[9]

The master bowman gathered a small army of 600 survivors from the town the dragon had destroyed. The ratio of archers to swordsmen was 2:3. How many archers followed the master bowman to the final battle?

Now our students have matured, graduating from the simple "three times as many" situations to full-fledged ratios. Without bar model diagrams (or similar pictorial methods), middle-school students find ratios an abstract and difficult subject. But a diagram makes it easy to see relationships.

The ratio tells how many parts (units) to draw for each group. "2:3" means two units of archers and three units of swordsmen.

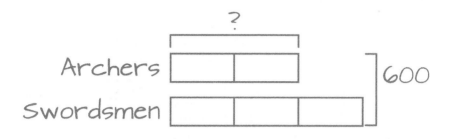

5 units = 600
1 unit = 600 ÷ 5 = 120
2 units = 120 × 2 = 240

Two hundred and forty archers followed the master bowman to battle.

[10]

> *The dwarfs rewarded their halfling friend with two chests of gold, silver, and small gems—6,000 pieces of treasure altogether. There were twice as many pieces of gold as there were gems. There were 600 more pieces of silver than gems. How much of each type of treasure did the halfling receive?*

We finish our chapter with another comparison problem, like the earlier one with the trolls. We hope our students have learned something in the interim, so this problem will require a few extra steps. For example, we want the amount for all three types of treasure, not just one.

We're comparing the number of pieces of gold, gems, and silver, so we will need three bars. The gems are compared to both the other treasures, so we put them in the middle. There are

twice as many (notice the ratio?) pieces of gold. And the pieces of silver match the gems plus 600 more:

As we did earlier in the troll problem, we first remove the "extra" pieces. This lets us work with the unknown units by themselves.

$$4 \text{ units} = 6000 - 600 = 5400$$
$$1 \text{ unit} = 5400 \div 4 = 1350$$
$$2 \text{ units} = 1350 \times 2 = 2700$$
$$1 \text{ unit} + 600 = 1950$$

And on complex problems, it's always a good idea to double-check. Do these numbers make the right total?

$$1350 + 2700 + 1950 = 6000$$

(Yes, it's important to check. I made a mental math error in the process of typing this chapter. I caught the mistake only because I took the time to add up my answers and see if they made sense.)

The halfling received 2,700 pieces of gold, 1,350 gems, and 1,950 pieces of silver.

Make Your Own Math

What sort of fractions do the people in your story world use? What do they measure or build? Do they use ratios to compare things, or to keep track of ingredients in a recipe? Have fun making up your math stories.

In the companion *Word Problems Student Workbook*, problems 11–13 come from Chapter 9 of this book. So we'll jump to #14 for our story problem prompts.

[14]

Write a scaling-up or scaling-down problem set in the world of a book or movie you enjoy.

There will be one group or part or thing that is some number of times as many or as big as something else. Or some fraction as much as the other amount.

Your problem has four numbers:

♦ The smaller amount.

♦ The larger amount.

♦ The scale factor.

♦ And the total of both amounts.

Tell any two of these numbers. Then ask your reader to figure out one (or both) of the others.

[15]

Write a problem that includes some type of ratio.

The groups or parts or things are related in a more complex way. Not just "twice as much" or "one-third as big," but using a ratio like 2:3.

How much information do you need to include so that your reader can figure it out?

[16]

Write a multistep problem set in the world of a book or movie you enjoy.

Have three or more parts to compare within your story. The

numbers may include the total amount and the sizes of the parts. Or instead of telling the size of a part directly, you might use a comparison (how much more or less) or ratio (how many times as much) to make a multistep puzzle-within-a-problem.

[17]

Write a transfer problem set in the world of a book or movie you enjoy.

Your problem needs an initial situation, and then one character gives some amount to another character, which sets up the final situation.

But be tricky about the information you give. Instead of telling the transferred amount directly, you might use a comparison or ratio to make a multistep puzzle-within-a-problem.

Go back and look at the goblin army problem for an example. How much information do you need to include so that your reader can figure it out?

[18]

Write a problem about sharing something. Use mixed numbers in the problem, or let the answer be a mixed number.

[19]

Write a problem that uses fractions with different denominators.

For an easier problem, make the denominators related to each other, like thirds and sixths. For a harder problem, use unrelated denominators.

[20]

Write a problem that involves a fraction of a fraction.

Have three or more groups or items, where at least one is

some fraction of another. Or tell a two-step problem where your character works with some fraction of a remaining part.

[21]

Write any kind of problem you like.

Conquer Monster Topics: Percents, Rates, and Proportional Reasoning

7

Word Problems Based on
Tales from Shakespeare

*Characters explore the heights and depths of human
nature in the Bard's comedies and tragedies.*

UPPER ELEMENTARY AND MIDDLE SCHOOL students face a
daunting array of "math monsters," tough topics like percents,
rates, and proportions, the kind of math that makes many children (and adults) feel like screaming. These topics are connected,
since they are all fractions in disguise, so students who have a
solid understanding of fraction math can build on that foundation to make sense of these difficult concepts.

At this stage, students move from thinking about multiplication as an operation (something they need to do) to multiplication as a relationship, a proportional connection between two
quantities. For example, if gasoline costs $5/gallon, that relation-
ship between dollars and gallons will remain the same for any
amount of gas: $10 for 2 gallons, $50 for 10 gallons, and so on.

Encourage students to play with ratio tables (introduced in
problem #2 below) for solving percent and rate problems, even
when they could work the problem without it. We want them to

grow comfortable with thinking in proportions, because this will ease their way into reasoning about functions when they move into algebra.

In this chapter, we'll explore a variety of math-monster word problems set in the world of Charles and Mary Lamb's *Tales from Shakespeare*.

As always, try your hand at solving the problems before reading my solutions.

[1] The Tempest

Caliban picked sugarcane and ground it for Miranda. He brought her 2.5 kg of sugar. Miranda used 325 g of the sugar to make cookies and 1.45 kg to make cakes. How much sugar did Miranda have left? Give the answer in kilograms.

[2] Midsummer Night's Dream

When the king of the fairies sent him to find a love-potion flower, Puck flew to a certain field. There were 200 flowers in the field. Only 2 of the flowers were the type Puck sought. What percentage of the flowers were love-potions?

[3] The Winter's Tale

The infant princess Perdita lay abandoned in a basket, along with jewelry valued at 750 gold pieces. A poor shepherd found her. He sold some of the jewels for 300 gold pieces, that he might provide a good home for his adopted daughter. What percentage of the treasure did he save to give to Perdita when she grew up?

[4] Much Ado about Nothing

When Don Pedro returned from the war, he brought a booty of 800 gold pieces. He wagered 3% of this treasure that he could trick Benedick and Beatrice into falling in love. How much money did Don Pedro wager?

[5] As You Like It

There were 60 men living with the banished duke in the forest of Arden. 55% of the men were wealthy friends and noblemen who had followed the duke. The rest were servants who came with their masters into exile. How many of the men in the forest were servants?

[6] Two Gentlemen of Verona

Faithless Proteus sent his page (fair Julia in disguise) to buy a fancy gift for the lady Sylvia, even though he knew Silvia did not love him. The gift cost 1,350 Lira plus a 3% tax. How much did Julia pay for this unwelcome gift?

[7] The Merchant of Venice

The merchant Antonio had three ships at sea. If the average profit from a merchant ship's journey was 2,500 ducats, how much did Antonio expect to gain when his ships came to dock?

[8] Cymbeline

After wicked Iachimo stole her bracelet, Imogen got lost in the forest. She stopped to drink at a small spring, where water flowed out between the rocks before running off to join a stream. 15 liters of water fell from the spring every 6 minutes. Find the rate of flow in liters per minute.

[9] King Lear

After being abandoned by his wicked daughters, King Lear wandered through the countryside. It took him 9 days to travel 45 km. At this rate, how long would it take him to travel 125 km to the earl of Kent's castle at Dover?

[10] Macbeth

The three witches heated their cauldron slowly, so the potion came almost to a boil. The potion produced 12 bubbles in 48 seconds. At this rate, how many bubbles would it make in one minute?

[11] All's Well That Ends Well

After Count Bertram rejected her as his wife, Helena traveled 750 km to Florence, Italy. If her average rate of travel was 30 km per day, how long did the journey take?

[12] Twelfth Night; Or, What You Will

Viola, disguised as the page Cesario, visited the lady Olivia to return her ring. They had tea in the garden, which had the shape of a rectangle 60 m long by 40 m wide. Find the ratio of the length to the width to the perimeter of the garden.

[13] Hamlet, Prince of Denmark

After a late-night game of cards, the ratio of Horatio's winnings to Hamlet's money was 2:5. Horatio won 24 guilders. How much money did Hamlet have?

[14] Othello

The general Othello led an army of 14,000 men to defend Cyprus against the invading Turks. 15% of the army were gentleman officers, and the rest were common soldiers. How many gentleman officers were under Othello's command?

[15] Pericles, Prince of Tyre

The lady Marina, captured by pirates and sold as a slave, earned money by teaching. Her master allowed her to keep 4% of her earnings to spend on herself. If Marina kept 6 talents, how much money did she earn?

[16] Romeo and Juliet

Romeo put on a mask and went uninvited to a party thrown by Lord Capulet. The number of men at the party was ⅝ the number of women. If there were 24 more women than men, how many people attended Lord Capulet's party?

Bar Models, Ratio Tables, and Answers

[1] The Tempest

Caliban picked sugarcane and ground it for Miranda. He brought her 2.5 kg of sugar. Miranda used 325 g of the sugar to make cookies and 1.45 kg to make cakes. How much sugar did Miranda have left? Give the answer in kilograms.

Percents are closely related to decimals, another fraction in disguise, which means we often make a quick review of decimal math before moving into percents. In this problem, we use measurement to give a context for decimal place value to the hundredths and thousandths.

Many students will want to work with the grams and kilograms separately, but encourage them to think in decimal fractions. One gram is $\frac{1}{1,000}$ of a kilogram, so 325 g is the same as 0.325 kg.

sugar

0.325	1.45	?

2.5

$$2.5 - (0.325 + 1.45) = ?$$
$$2.5 - 1.775 = 0.725$$

Miranda had 0.725 kg of sugar left.

[2] Midsummer Night's Dream

When the king of the fairies sent him to find a love-potion flower, Puck flew to a certain field. There were 200 flowers in the field. Only 2 of the flowers were the type Puck sought. What percentage of the flowers were love-potions?

We introduce percents with the keyword definition: "per cent" means "per hundred." Thus, the first way children learn to handle percent calculations is to convert a given fraction, decimal, or ratio into hundredths.

In these situations, a ratio table is often more useful than a bar model. A *ratio table* is a simple chart of numbers in a given proportion, which we can modify like equivalent fractions until we find the ratio that solves our problem.

love potions	2	?
all flowers	200	100

÷2

$$2:200 = 1:100$$

One flower of every hundred was a love-potion flower. Therefore, 1% of the flowers were love-potions.

[3] The Winter's Tale

The infant princess Perdita lay abandoned in a basket, along with jewelry valued at 750 gold pieces. A poor shepherd found her. He sold some of the jewels for 300 gold pieces, that he might provide a good home for his adopted daughter. What percentage of the treasure did he save to give to Perdita when she grew up?

A *benchmark value* is a number that is easy to work with, especially when doing mental math. The key benchmark values for percents are:

- ◆ 100% = the whole thing
- ◆ 50% = half of 100%
- ◆ 25% = half of 50%, or ¼ of the whole amount
- ◆ 10% = ⅒ of the whole amount
- ◆ 5% = half of 10%
- ◆ 1% = ⅒ of 10%, or ⅟₁₀₀ of the whole amount

10,000	20,000	30,000	40,000	50,000	60,000	70,000	80,000	90,000
1,000	2,000	3,000	4,000	5,000	6,000	7,000	8,000	9,000
100	200	300	400	500	600	700	800	900
10	20	30	40	50	60	70	80	90
1	2	3	4	5	6	7	8	9
0.1	0.2	0.3	0.4	0.5	0.6	0.7	0.8	0.9
0.01	0.02	0.03	0.04	0.05	0.06	0.07	0.08	0.09

A *Powers of Ten* (or *Orders of Magnitude*) number chart helps students to reason about the ten-of-this and one-tenth-of-that relationships that define our decimal number system.

Practice Benchmark Numbers with a Powers of Ten Chart

Start at any number.

When you move up one row, you find ten times the number. Moving up two rows gives you one hundred times the number, and so on.

When you move down one row, you find ¹⁄₁₀ of the number. Moving down two rows gives you ¹⁄₁₀₀, and so on.

- ◆ If your original number is 400, what is 10% (or ¹⁄₁₀) of that number?

- ◆ If 1% (or ¹⁄₁₀₀) is 60, what must be the whole amount (100%)?

Have your children help you draw a Powers of Ten chart on a large sheet of paper. Extend it to include as many rows as you like.

- ◆ Ask questions that require students to think about moving up and down the powers of ten.

- ◆ Encourage them to pose questions for you, too. Can they stump the adult?

- ◆ Think about the numbers between the columns. For example: 10% of 50 is 5, and 10% of 60 is 6. So what must be 10% of 57?

To solve *The Winter's Tale*, we might find the percentage of the 300 gold pieces, and then subtract that from 100 to find the number we want. Or we could find the value of the remaining treasure in gold pieces, and then calculate the percentage directly.

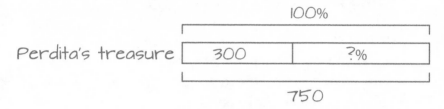

$$750 - 300 = 450$$
$$450 = ?\% \text{ of } 750$$

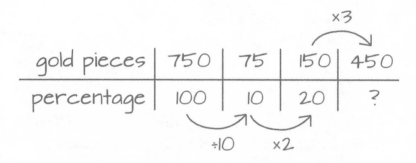

100% = 750 gold pieces
10% of 750 = ¹⁄₁₀ of 750 = 75
20% of 750 = 2 × 75 = 150
450 = 3 × 150 = ?% of 750
3 × 20% = 60%

The shepherd saved 60% of the treasure for Perdita.

[4] Much Ado about Nothing

When Don Pedro returned from the war, he brought a booty of 800 gold pieces. He wagered 3% of this treasure that he could trick Benedick and Beatrice into falling in love. How much money did Don Pedro wager?

We can use a ratio table to convert 3% (3:100) into the equivalent ratio N:800:

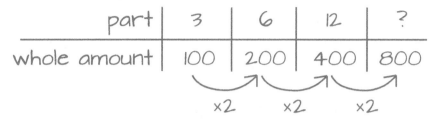

Or we can think in benchmark numbers:

$$100\% = 800 \text{ gold pieces}$$
$$1\% \text{ of } 800 = \tfrac{1}{100} \text{ of } 800 = 8$$
$$3\% \text{ of } 800 = 8 \times 3 = 24$$

Whichever way we calculate it, Don Pedro wagered 24 gold pieces.

[5] As You Like It

There were 60 men living with the banished duke in the forest of Arden. 55% of the men were wealthy friends and noblemen who had followed the duke. The rest were servants who came with their masters into exile. How many of the men in the forest were servants?

As in *The Winter's Tale* above, here we are given information about one part of the whole amount and asked to find the other part, the *complement*—the part that *completes* the total amount. Two percentage values that together make 100% are complements of each other.

```
                          100%
         ┌─────────────────────────────────────┐
exiles   │        ?        │        55%         │
         └─────────────────────────────────────┘
                           60
```

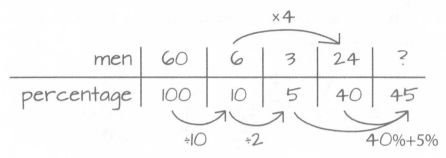

The complement of 55% is 45%, so we need to figure out how many men are 45% of the 60 exiles.

$$45\% \text{ of } 60 = ?$$
$$10\% \text{ of } 60 = \tfrac{1}{10} \text{ of } 60 = 6$$
$$5\% \text{ of } 60 = 6 \div 2 = 3$$
$$45\% \text{ of } 60 = (4 \times 6) + 3 = 27$$

There were 27 servants in the banished duke's group.

[6] Two Gentlemen of Verona

Faithless Proteus sent his page (fair Julia in disguise) to buy a fancy gift for the lady Sylvia, even though he knew Silvia did not love him. The gift cost 1,350 lire plus a 3% tax. How much did Julia pay for this unwelcome gift?

Tax is calculated by counting the purchase price as 100%, and then adding an extra cost that is some percentage of that purchase price. That means Julia ends up paying more than 100% for her purchase, which may confuse some students.

100%

| cost | 1350 | 3% |

?

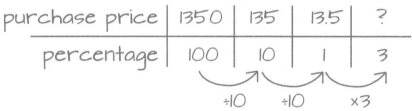

purchase price	1350	135	13.5	?
percentage	100	10	1	3

÷10   ÷10   ×3

100% = 1350
1% of 1350 = ¹⁄₁₀₀ of 1350 = 13.50
3% of 1350 = 13.50 × 3 = 40.50
1350 + 40.50 = 1390.50

Julia paid 1,390.50 lire for the gift.

[7] The Merchant of Venice

The merchant Antonio had three ships at sea. If the average profit from a merchant ship's journey was 2,500 ducats, how much did Antonio expect to gain when his ships came to dock?

The real world has jagged edges. Even when we try to pour the juice fairly, no two children get exactly the same amount. We drive to town, and our speed goes up and down in response to road conditions. Merchant ships have varying degrees of success in trading.

Averages smooth out the roughness of real life. An average

asks, "If all things were equal, then how much would each value be?"

Averages are estimated this-per-that rates, where we assume every "that" gets the same amount of "this." In elementary and middle school math problems, all rates are steady, average rates. In high school, the rates may vary—but only in steady, average ways. Only in calculus will students gain the tools to handle real-world variation.

The average rate of return could not predict Antonio's exact profit, but it gave him a way to calculate anticipated earnings.

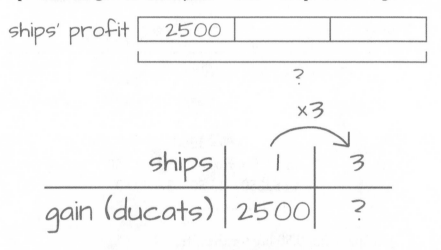

One unit = 2500
Three units = 2500 × 3 = 7500

Antonio expected to gain approximately 7,500 ducats.

[8] Cymbeline

After wicked Iachimo stole her bracelet, Imogen got lost in the forest. She stopped to drink at a small spring, where water flowed out between the rocks before running off to join a stream. 15 liters of water fell from the spring every 6 minutes. Find the rate of flow in liters per minute.

We assume the spring produces water at a steady rate, the same amount every minute. In six minutes, it produced 15 liters of water. Therefore, in one minute, it must produce ⅙ of that amount.

liters	15	5	?
minutes	6	2	1

÷3   ÷2

$$^{15\ \text{liters}}/_{6\ \text{min}} = (15 \div 6)\ ^{\text{liters}}/_{\text{min}} = 2.5$$

The spring produced 2.5 liters per minute.

[9] King Lear

After being abandoned by his wicked daughters, King Lear wandered through the countryside. It took him 9 days to travel 45 km. At this rate, how long would it take him to travel 125 km to the earl of Kent's castle at Dover?

As in Cymbeline's problem, when we know any rate, we can simplify it to a *unit rate*—the rate with one in the denominator. If King Lear travels 45 km in 9 days, he must travel ⅑ of that distance per single day.

Unit rates are useful in solving problems because when you know the rate per single day, it's easy to multiply by as many days as you need, or to divide the daily amount into whatever distance you plan to travel.

$$^{45\ \text{km}}/_{9\ \text{days}} = (45 \div 9)\ ^{\text{km}}/_{\text{day}} = 5$$
$$125 \div 5 = 25$$

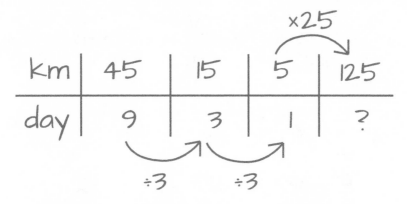

King Lear needed 25 days to travel to the earl's castle.

[10] Macbeth

The three witches heated their cauldron slowly, so the potion came almost to a boil. The potion produced 12 bubbles in 48 seconds. At this rate, how many bubbles would it make in one minute?

One minute is sixty seconds, so if we can find the unit rate of bubbles per second, then we'll be able to multiply that and figure out how many bubbles there are per minute.

bubbles	12	3	1	0.25	?
seconds	48	12	4	1	60

$$\frac{12\ \text{bubbles}}{48\ \text{seconds}} = (12 \div 48)\ \text{bubbles}/\text{sec} = 0.25$$
$$0.25 \times 60 = 15$$

The slowly simmering potion made 15 bubbles per minute.

[11] All's Well That Ends Well

*After Count Bertram rejected her as his wife, Helena traveled
750 km to Florence, Italy. If her average rate of travel was
30 km per day, how long did the journey take?*

In this problem, we are given the unit rate ($^{km}/_{day}$), but what
shall we do with it? How can your students know when they
need to multiply by a rate and when to divide?

Math is not about following someone else's rules, but about
making sense for ourselves. Therefore, we need to help children
make sense of rate problems, rather than just relying on a memorized formula like "rate × time = distance."

A ratio table is one way to make sense. We start with the
unit rate and make equivalent ratios until we get to the distance
needed.

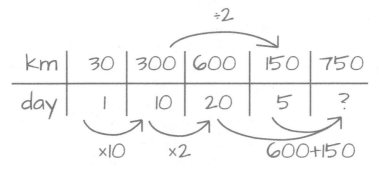

Or we can think about traveling a certain distance per day,
and again the next day, and again. How many chunks of 30 km
make up the whole journey? Does it seem reasonable that the
total distance divided by the amount traveled per day should
give the answer we need?

$$30 \,^{km}/_{day} \times (\text{how many days?}) = 750 \text{ km}$$
$$750 \div 30 = 25$$

Helena took 25 days to travel that distance.

[12] Twelfth Night; Or, What You Will

Viola, disguised as the page Cesario, visited the lady Olivia to return her ring. They had tea in the garden, which had the shape of a rectangle 60 m long by 40 m wide. Find the ratio of the length to the width to the perimeter of the garden.

By this time, students should be familiar with simple, two-part ratios. But we can use ratios to compare any number of items. For example, a recipe can be understood as an extended ratio of ingredients. And any ratio, like any fraction, can be put into simplest form (lowest terms).

To find the perimeter, imagine starting at one corner and walking the length and then the width of the garden. That gets you halfway around, so you must walk the same distance again to find the whole perimeter.

$$\text{Length + width} = 60 + 40 = 100$$
$$\text{Perimeter} = 100 \times 2 = 200$$
$$\text{Ratio} = 60{:}40{:}200 = 3{:}2{:}10$$

The ratio of length to width to perimeter is 3:2:10.

length	60	6	3
width	40	4	2
perimeter	200	20	10

÷10   ÷2

[13] Hamlet, Prince of Denmark

After a late-night game of cards, the ratio of Horatio's winnings to Hamlet's money was 2:5. Horatio won 24 guilders. How much money did Hamlet have?

Because a ratio can be written as a fraction, and because we know that every equivalent fraction has terms related in the same way as the original ratio, we can solve ratio problems by writing a *proportion*. Think of a proportion as an extremely simplified ratio table with just two entries, two equal fractions.

$$\text{Horatio's winnings : Hamlet's money} = \tfrac{2}{5} = \tfrac{24}{?}$$

To solve the proportion, we must find the missing number that will make the equation true. The numerator of the first fraction is multiplied by twelve in the second fraction, so we know the denominator must go up by the same multiple.

$$\text{Horatio's winnings} \qquad \overset{\times 12}{\frac{2}{5}} = \frac{24}{?}$$

Hamlet's money

$$\tfrac{2}{5} = \tfrac{24}{60}$$

Hamlet had 60 guilders.

[14] Othello

The general Othello led an army of 14,000 men to defend Cyprus against the invading Turks. 15% of the army were gentleman officers, and the rest were common soldiers. How many gentleman officers were under Othello's command?

In addition to ratios, we can solve percent problems by setting up a proportion. The part we are interested in compares to the *base* (the amount that counts as 100%) as the percentage value compares to 100.

$$\text{part}/\text{base} = \text{percentage}/100$$

The base is the amount to which we are comparing our part. The base is not necessarily "all that there is." It's not a "whole" in that sense. It's just the basis for our comparison. In some problems, the part could even be larger than the base, creating a percentage greater than 100%.

Math vocabulary is not intuitive. Even in high school, you will probably still be reminding students that the base is "the number being treated as 100% in this comparison." Just as you still remind them that the denominator is the bottom number of a fraction or that the perimeter is the distance around the "rim" of a shape.

$$\text{gentlemen officers} \quad \frac{?}{} \quad = \quad \frac{15}{}$$

$$\text{whole army} \quad \frac{}{14,000} \quad = \quad \frac{}{100}$$

$\times 140$

In this problem, the part we are trying to find is the number of gentleman officers, and the base is the whole army. Once the student has identified the part and the base, the most challenging thinking is done. The rest of the solution is merely crunching through the arithmetic.

$$^?/_{14,000} = {}^{15}/_{100}$$
$$100 \times 140 = 14,000$$
$$15 \times 140 = \; ? = 2,100$$

There were 2,100 gentleman officers under Othello's command.

[15] Pericles, Prince of Tyre

The lady Marina, captured by pirates and sold as a slave, earned money by teaching. Her master allowed her to keep 4% of her earnings to spend on herself. If Marina kept 6 talents, how much money did she earn?

In this story, we know the part of her earnings that Marina keeps, and we know what percentage that part is of the whole amount. We need to find the base, the whole amount she earned by teaching.

$$^{\text{part}}/_{\text{base}} = {}^{\text{percentage}}/_{100}$$
$$^6/_? = {}^4/_{100}$$

But there is no easy conversion between these fractions. What can our students do next?

You may be tempted to solve this equation by cross-multiplying but be warned: That will only cause your children trouble in the future. "Cross-multiply" means too many different things to different people, confusing students, leading to wrong answers, throwing a stick into the web of mathematical understanding.

(For more on the trouble with cross-multiplication and other math shortcuts, download Tina Cardone's free booklet *Nix the Tricks*.[†])

Instead of teaching a rule that will eventually be misremembered, let's apply some sense-making logic. Look again at the proportion. What do you notice? If this was a ratio table, what equivalent ratios might you use to figure it out?

$$\frac{\text{Marina's part}}{\text{base earnings}} \quad \frac{6}{?} = \frac{4}{100} = \frac{1}{25}$$

$$\times 6 \qquad \div 4$$

$$\frac{6}{?} = \frac{4}{100} = \frac{1}{25}$$
$$25 \times 6 = ? = 150$$

Marina earned 150 talents for her teaching.

[16] Romeo and Juliet

Romeo put on a mask and went uninvited to a party thrown by Lord Capulet. The number of men at the party was ⅝ the number of women. If there were 24 more women than men, how many people attended Lord Capulet's party?

Problems like this demonstrate the close connection between fractions and ratios. Any two-part ratio may be written as a frac-

---

† *nixthetricks.com*

tion, and any fraction may be read as the ratio of numerator to denominator.

To say that the number of men was ⅝ the number of women is the same as saying the ratio of men to women is 5:8. Examine the bar diagram. Can you see how both statements are true?

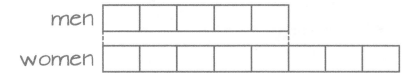

Now we can add the numbers to our diagram. There are 24 more women than men, and we want to find the total number of people at the party.

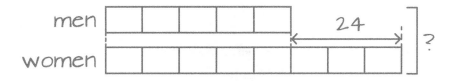

3 units = 24
1 unit = 24 ÷ 3 = 8
13 units = 8 × 13 = 104

There were 104 people at Lord Capulet's party.

Make Your Own Math

Where in your favorite story world would the characters use rates, ratios or percents? Do they measure speed, or buy steampunk gears by the pound, or pay a worker by the hour? When might they need to compare a part to a whole thing?

What stories will you create?

[17]

Write a scaling-up or scaling-down problem using a ratio, set in the world of a book or movie you enjoy.

Your problem has five numbers:

- ♦ The smaller amount, and the larger amount.
- ♦ The difference between them, and the total of both amounts.
- ♦ The ratio comparison.

Tell the ratio and any *one* of the other numbers. (You may tell more than one number, if you wish.) Then ask your reader to figure out whichever of the remaining numbers you like.

[18]

Write a problem that uses percents. Where in your story world would a character work with percent calculations? Think about comparing parts to a whole group, making a wager, dividing the spoils from a raid, and things like that.

The numbers in a percent problem may include whatever counts as the base (100%), the part of that thing we are interested in, and the percentage value. Or you might use the complement, which is the percentage value for the part we are *not* interested in.

Tell two of the percentage (or the complement), the part, and the base. Then ask your reader to figure out the missing number.

[19]

Write a multistep problem set in the world of a book or movie you enjoy. Have two or more parts to compare within your story.

The numbers may include the total amount and the sizes of the parts. But instead of telling the size of a part directly, use a fraction, ratio, or percent comparison to make a multistep puzzle-within-a-problem.

How much information do you need to include so that your reader can figure it out?

[20]

Write a rate problem, where your characters do something or go somewhere at a this-per-that rate, or they have some amount of this or of that which must be processed or poured out or filled up or traveled.

Your problem has three numbers:

♦ How fast your characters work or travel (the rate).

♦ How much they get done.

♦ How long it takes them.

Tell any two of these numbers. Then ask your reader to figure out the third.

[21]

Write any kind of problem you like.

Move Toward Algebra

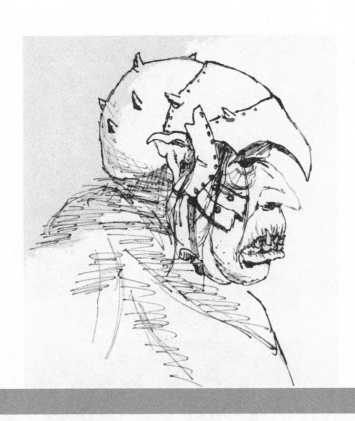

Challenge Problems

HAVE YOU MASTERED BAR MODEL diagrams yet? See what you can do with these challenging, multistep story problems. [I've numbered this chapter to match the *Word Problems Student Workbook*, which includes two problems from Chapter 9.]

[3] Space Opera Repairs

The rogue space smuggler Peyton and her first mate Spider performed maintenance work on their spaceship. Peyton spent ⅗ of her money upgrading the hyperspace motivator. She spent ¾ of the remainder to install a new blaster cannon. If Peyton spent 450 money chips altogether, how much money did she have left?

[4] The Dragon's Collection

A captive princess worked all afternoon cleaning and organizing part of Dragon Philip's treasure. ¼ of the items she sorted were jewelry. 60% of the remainder were potions, and the rest were magic swords. If there were 48 magic swords, how many pieces of Dragon Philip's treasure did she sort in all?

[5] A Fantasy World at War

Knowing the goblin horde would soon be upon them, Queen Emily placed 1,500 guards atop the outer wall at the canyon entrance that protected her fortress. 300 of the queen's guards were swordsmen and the rest were archers. What percentage fewer swordsmen were there than archers?

Solution 3: Space Opera Repairs

Inspired by the world of George Lucas's original *Star Wars:*

The rogue space smuggler Peyton and her first mate Spider performed maintenance work on their spaceship. Peyton spent ⅗ of her money upgrading the hyperspace motivator. She spent ¾ of the remainder to install a new blaster cannon. If Peyton spent 450 money chips altogether, how much money did she have left?

How can we teach our students to solve complex, multistep word problems? Depending on how one counts, our problem would take four or five steps to solve. One might approach it with algebra, writing a two-equation system like this:

$$x + 450 = y$$
$$\text{and}$$
$$y - \tfrac{3}{5}\,y - (\tfrac{3}{4} \times \tfrac{2}{5})\,y = x$$

... and then simplify the equations and solve for *x*.

But I took the basic structure of this problem from a puzzle in *Singapore Primary Mathematics 5B*. Fifth-grade students have not learned algebra yet, and many adults have forgotten most of what we ever knew. So let's apply the magic of a bar model diagram.

We start with a bar representing the money our smuggler had to start with.

Money chips

Three-fifths of the money went to upgrade the hyperspace motivator. This is easy to show by dividing the bar into five sections and marking three of them as spent.

Money chips

A Part Becomes a "Whole"

The smuggler used three-fourths of the remainder to install a blaster cannon. The words "of the remainder" are easy to overlook, so be careful. These words show that we have a new "whole thing." We draw a new bar below the original.

Money chips

On this new bar, three-fourths went for the blaster cannon, and the rest is still in the smuggler's pocket. We show this by dividing the bar into four pieces and mark three of the pieces as used.

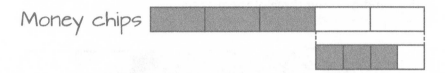

Money chips

Simplify to a Single Unknown Unit

We're almost done. The unmarked part of our "remaining money" bar is the money our smuggler has at the end of the story—what we are trying to find. We know that the money she spent adds up to 450 money chips. There are units of spent money in our diagram, but we have a problem:

The units are *not* the same size.

The units on the original bar model are different from the units on the "remaining money" bar. If the units were all the same size, we could divide the total amount spent by the number of units.

Is there any way we can adjust our diagram to create same-size units?

Notice that each small unit is half the size of a larger, original unit. That means we can divide each of those original units in half, and then they'll match the small units on the second bar. This is the bar diagram equivalent of finding a common denominator for fractions.

Money chips

(Experienced students may notice the relationship between the units before drawing the "remaining money" bar. They can divide the larger units in half at that point, making it possible to mark off the money spent for the blaster cannon without drawing a second bar. There is almost always more than one way to think through a story problem diagram.)

The rogue space smuggler spent nine of these smaller units of money for a total of 450 money chips. If we merge our two bars back into one, it's easy to see.

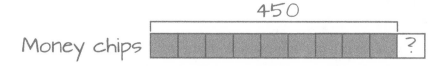

As soon as we can connect a unit (or a set of same-size units) with a number, the difficult task of thinking through the problem is done. Now all we need is basic arithmetic.

$$9 \text{ units} = 450$$
$$1 \text{ unit} = 450 \div 9 = 50$$

And one unit is the money left after finishing the repairs. In the end, our smuggler has only fifty money chips to her name.

Solution 4: The Dragon's Collection

Inspired by the world of Patricia Wrede's *Dealing with Dragons:*

A captive princess worked all afternoon cleaning and organizing part of Dragon Philip's treasure. ¼ of the items she sorted were jewelry. 60% of the remainder were potions, and the rest were magic swords. If there were 48 magic swords, how many pieces of Dragon Philip's treasure did she sort in all?

One might approach it with algebra, writing an equation like this:

$$x - [¼\, x + (0.6 × ¾)\, x] = 48$$

... and solve for the variable x. But again, this problem is for students who haven't learned algebra.

Instead, we start with a long bar representing the part of Dragon Philip's treasure that the princess sorted, as if she laid it in a row on the floor of the dragon's cave.

Treasure

One-fourth of the treasure was jewelry. That's easy to show by dividing the bar into four sections and marking one of them to stand for the jewelry.

Treasure

Sixty percent "of the remainder" were potions. That means we're treating the non-jewelry treasure as if the jewelry did not exist. We draw a new bar below the original.

Treasure

On this new bar, sixty percent stands for the potions, and the rest is the swords. Since $60\% = {}^{60}\!/\!_{100} = {}^{6}\!/\!_{10} = ⅗$, we can divide the bar into five pieces and mark three of the pieces as potions. The rest stands for the forty-eight magic swords.

We could simplify to a single unit by dividing the original units into five pieces each and the remainder units into three

Treasure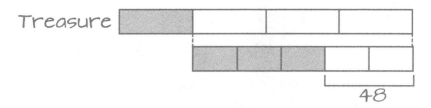

pieces each. Then each bit would represent one-twentieth of the treasure.

Twentieths? Such small fractional pieces are hard to draw. So we'll work with each bar separately.

$$2 \text{ small units} = 48$$
$$1 \text{ small unit} = 48 \div 2 = 24$$
$$5 \text{ small units} = 5 \times 24 = 120$$

There were 120 pieces of treasure, not counting the jewelry. Once we know this total, we are well on our way to answering the original question. Notice that the potions and swords together make up three units of the original bar, which represented the whole treasure.

$$3 \text{ large units} = 120$$
$$1 \text{ large unit} = 120 \div 3 = 40$$
$$4 \text{ large units} = 4 \times 40 = 160$$

So the princess sorted and cleaned 160 pieces of Dragon Philip's treasure. That's a busy afternoon's worth of work.

Solution 5: A Fantasy World at War

Inspired by the world of J. R. R. Tolkien's *The Lord of the Rings:*

> *Knowing the goblin horde would soon be upon them, the king placed 1,500 guards atop the outer wall at the canyon entrance that protected his fortress. 300 of the guards were swordsmen and the rest were archers. What percentage fewer swordsmen were there than archers?*

Percents: The Search for 100%

Percents are one of the math monsters, the toughest topics of arithmetic. The most important step in solving any percent problem is to figure out what quantity is being treated as the *base,* the thing that equals 100%. Everything else in the problem is measured in comparison to the base.

Notice the variety of phrases we can use to signal 100%...

_____ *% of* _____
100% = whatever comes after the word "of."

_____ *as a percentage of* _____
100% = whatever comes after the word "of."

_____ *% comparison: more/less/greater/fewer than* _____
100% = whatever comes after the word "than."

_____ *% increase, decrease, discount, or rebate*
100% = the original amount or list price.

_____ *% gain, profit, or loss*
100% = the original cost of the item to the seller.

_____ *% sales tax, down payment, deposit, or commission*
100% = the final cost of the item to the buyer.

_____% interest

100% = the principal of the account or loan.

_____% raise

100% = the worker's previous wages.

_____% income tax

100% = the person's annual income, or a portion of that income in the case of a progressive tax.

_____% enlargement or reduction

100% = (usually) the linear dimensions of the original photo or drawing.

A solution containing _____% of some chemical

100% = the volume (or mass, depending on the problem) of the entire solution, *not* the chemical listed after the word "of."

… and there may be other situations I've missed. Is it any surprise that many people struggle with percent problems?

Of all these percent situations, the most difficult for students to analyze—and sometimes even for teachers—is the percent comparison. Some quantity is a certain percentage more or less than another.

Which brings us back to our queen and her guards:

Knowing the goblin horde would soon be upon them, the king placed 1,500 guards atop the outer wall at the canyon entrance that protected his fortress. 300 of the guards were swordsmen and the rest were archers. What percentage percent fewer swordsmen were there than archers?

The most common mistake with a problem like this is to count all the guards as 100%. So a student will figure 20% of the guards are swordsmen and 80% are archers. Thus there must be 80 – 20 = 60% fewer swordsmen than archers, right?

Wrong!

In this problem, the words "than archers" show we are comparing the swordsmen to *the number of archers alone*. Whatever we are comparing to, that is what we treat as 100%. So in this problem, the base is the 1,200 archers.

This problem has one more potential stumbling block. Remember that the basic percent proportion is:

$$\text{part}/\text{base} = \text{percentage}/100$$

Careless students will ignore the word "fewer" and use the number of swordsmen as the part in their proportion. But "fewer" tells us to focus on the *difference* between the swordsmen and the archers. We need to find that difference by subtracting before we are ready to calculate the percentage.

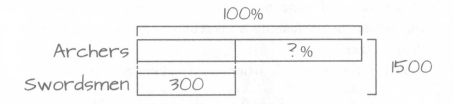

1500 – 300 = 1200 archers
1200 – 300 = 900 fewer swordsmen

$$\text{part}/\text{base} = \text{percentage}/100$$

$$900/1200 = \text{percentage}/100$$

$$\tfrac{3}{4} = \text{percentage}/100$$

$$\tfrac{3}{4} \times 100 = \text{percentage} = 75$$

There were 75% fewer swordsmen than archers.

Teaching Tips

♦ Help your students learn to recognize 100% in all its disguises and to think their way through those tricky "than" comparisons.

♦ Be careful to identify the base in each percent problem.

♦ Watch out for parts that require subtraction, like how many "more" or "fewer."

♦ Then set up a percent proportion. Most of the time, your solution will fall into place.

Percents are indeed a mathematical Jabberwock, with jaws that bite and claws that catch many a careless student. But even a monster like percents will fall to a well-swung vorpal sword. With a good understanding of fractions and plenty of practice applying common sense, your students will be able to forge their way fearlessly through the tulgey wood of arithmetic.

Transition to Algebra

Once upon a time, the goal of teaching elementary arithmetic was exactly that. Teachers trained children to work with numbers so that as adults they would be able to work with numbers. But nowadays, we all carry in our pocket or purse a little computer that can easily do any difficult calculation.

Today, we teach children to work with numbers because this helps them to *think* about numbers. We want them to be able to reason with numbers, recognizing number relationships, noticing the patterns they make, using those patterns to solve problems.

In other words, we want them to master algebra.

Algebra is generalized arithmetic. It's a mathematical language that helps us think about number patterns directly.

With arithmetic, we can know that 2 + 3 = 5. And that 1 + 4 = 5, and also −11 + 6. We use arithmetic to reason about specific numbers.

But with algebra, we can write A + B = 5. In one expression, we capture all the pairs of numbers with that sum, whether they are whole numbers, fractions, decimals, positive, negative, or even complex numbers with irrational parts. Algebra lets us reason about all these numbers at the same time, using a single equation.

As a problem-solving tool, algebra has the advantage of efficiency, packing information tightly into a few marks on the page, helping us get to an answer quickly. And algebra is flexible, applying to almost any situation involving numbers, adapting itself to many uses in science or higher math.

But the very thing that makes algebra so efficient and flexible also makes it difficult for students to learn—namely, its abstraction. "Math with letters" offers children no hook to catch their imagination.

Word Algebra

We can bridge the chasm of abstraction by teaching students to use words as variables. Word algebra helps students think about the meaning of a math equation and see how it applies to the situation at hand.

For example, let's look back at this puzzle from Chapter 1:

There are 21 girls in a class. There are 3 times as many girls as boys. How many boys are in the class?

This problem has three unknown quantities. We can use the words *class, girls,* and *boys* to represent the number of children in each group, letting us summarize the story in equations.

$$class = girls + boys$$

$$girls = 21 = 3 \times boys$$
$$boys = ?$$

Now any student who knows the inverse relationship between multiplication and division can solve the riddle.

$$21 = 3 \times boys$$
$$boys = 21 \div 3 = 7$$

When we use words as variables, we enable students to understand the meaning of equations, so they can apply their own common sense to the mathematical situation.

As word algebra equations grow more complex, students must learn to use a few basic but important rules.

The Rule of Inverse Operations

If a mathematical operation has been used on some quantity, you can "undo" it by applying the inverse of that operation. For example, if you know that *number* + 5 = 8, then you can undo the addition by subtracting five.

The Balance Rule

If you want to use a new mathematical operation on one side of an equation, you must also do the same thing to the other side, to keep the equation in balance. So in the example *number* + 5 = 8, if you want to subtract five from the "*number* + 5" side of the equal sign, you must also subtract five from the 8.

The Rule of Substitution

If one quantity is equal to another quantity, then it may be substituted for the other quantity in any equation. Thus, once you have determined that *number* = 3 in this particular problem, you can then use that information in any other equation within the same problem.

The Distributive Property

If you want to add (or subtract, or multiply) *a group of quantities,* usually set apart from the rest of the equation by parentheses, you must think very carefully about what the math symbols mean. In most cases, you need to add (or subtract, or multiply) everything within that group, one at a time. And you must do this very carefully, especially when subtracting, because this is one of the easiest places to make a mistake in algebra.

For example, if you double the number of children in a group, you must double both the number of boys and the number of girls. Or if a group of children leaves, you must subtract the number of boys and also subtract the number of girls in that group.

$$2 \times (\textit{boys} + \textit{girls}) = (2 \times \textit{boys}) + (2 \times \textit{girls})$$
and
$$\textit{people} - (\textit{boys} + \textit{girls}) = \textit{people} - \textit{boys} - \textit{girls}$$

After students grow comfortable with the idea of word algebra, they will gradually shorten the words to just initials. This natural development mirrors math history. When Muhammad ibn Musa al-Khwarizmi wrote the first algebra book, he used no equations at all, writing everything in words. Over the centuries, mathematicians adopted shorthand for the written words, gradually creating the symbolic algebra we use today.

Let's try a few word algebra problems...

Word Problems with King Arthur

A medieval king has adventures with his knights and a wizard.

[6]

> *King Arthur gathered several of his knights to send them on a quest. He had a bag of silver coins to give them for provisions along the way. If he gave each knight 8 coins, he would have 4 coins left in his bag. But if he gave only 5 coins to each knight, he would have 40 coins left. How many knights were going on the quest?*

[7]

> *King Arthur had a small chest of jewels in his treasury, gifts from the many foreign dignitaries who visited his court. 1/3 of the stones were rubies, 1/9 were emeralds, and 1/5 of the remainder were diamonds. If there were 25 diamonds, how many jewels did King Arthur have in the chest?*

[8]

> *King Arthur and Merlin went shopping at the Camelot Fair. They both had the same number of coins. Then Arthur spent 50 coins on new tapestries for the walls in his castle, to keep down the winter chill. And Merlin spent 1/3 of his money on new books and scrolls to study.*
>
> *When they compared their purses at the end of the day, the ratio of King Arthur's money to Merlin's was 5:4. How many coins did they each have at the beginning of the day?*

Word Algebra and Answers

[6]

> *King Arthur gathered several of his knights to send them on a quest. He had a bag of silver coins to give them for provisions along the way. If he gave each knight 8 coins, he would have 4 coins left in his bag. But if he gave only 5 coins to each knight, he would have 40 coins left. How many knights were going on the quest?*

Word algebra lacks the visual clarity of a bar model diagram. It will take more writing to explain what we could have seen at a glance in the model. On the other hand, word algebra lets us approach problems like this, which students may find difficult to put into a model.

We have two unknown quantities: the number of knights going on the quest, and the total number of coins in the bag. We can use the words *knights* and *moneybag* to represent these amounts.

When King Arthur gives each knight eight coins, that removes $8 \times knights$ coins from his bag and leaves him with four coins.

$$moneybag - (8 \times knights) = 4$$

But if the king gives each knight only five coins, we have a different equation.

$$moneybag - (5 \times knights) = 40$$

Those are the two facts we know. And our goal is to discover the number of knights.

$$knights = \;?$$

What can we do with this information? How can we apply common sense to this situation?

Let's start by looking more closely at our two equations, writing them next to each other to look for a pattern.

$$moneybag - (8 \times knights) = 4$$
$$moneybag - (5 \times knights) = 40$$

Hmm. When King Arthur gave three additional coins to each knight, he ended up with 36 fewer coins in his bag. That gives us one more equation to work with, which turns out to be the key:

$$3 \times knights = 36$$
$$knights = 36 \div 3 = 12$$

King Arthur was sending twelve knights on the quest.

[7]

King Arthur had a small chest of jewels in his treasury, gifts from the many foreign dignitaries who visited his court. $\frac{1}{3}$ of the stones were rubies, $\frac{1}{9}$ were emeralds, and $\frac{1}{5}$ of the remainder were diamonds. If there were 25 diamonds, how many jewels did King Arthur have in the chest?

Earlier in this chapter, we saw how bar model diagrams help us make sense of fraction-of-a-remaining-part problems. We can also solve them with word algebra, though the sheer quantity of writing may intimidate students. You might want to scribe for your students, letting them simply talk about what they see in the problem while you write down whatever they say.

Keep in mind that we want our students to be reasoners, not rote rule-followers. So we need to walk through the problem in reasonable steps, making sense of it as we go.

This problem offers us a gaggle of unknown quantities. We have the total number of jewels and the numbers of certain types: rubies, emeralds, and diamonds. Using each name to represent that number, we can list the following bits of information:

$$\tfrac{1}{3} \text{ of } jewels = rubies$$
$$\tfrac{1}{9} \text{ of } jewels = emeralds$$
$$\tfrac{1}{5} \text{ of } [jewels - (rubies + emeralds)] = diamonds$$
$$diamonds = 25$$

And we want to find the total number of jewels, so:

$$jewels = ?$$

What a mess! Here's a tip: When we find ourselves with a slew of variables, the Rule of Substitution comes to our rescue. Look for things to substitute that will reduce the number of variables in your problem.

For example, we already know how many diamonds there are. So we can substitute the number 25 everywhere we see *diamonds*.

$$\tfrac{1}{5} \text{ of } [jewels - (rubies + emeralds)] = 25$$

And we don't need to know what's inside the brackets to apply our knowledge of fractions. If $\tfrac{1}{5}$ of something is 25, then the whole amount must be five times as much.

$$\tfrac{1}{5} \text{ of } [some\ number\ I'll\ figure\ out\ later] = 25$$
$$[some\ number\ I'll\ figure\ out\ later] = 25 \times 5 = 125$$
$$jewels - (rubies + emeralds) = 125$$

Now before we go any farther, we'd better apply the Distributive Property to subtract the rubies and emeralds. But, carefully! If students think about what it really means to subtract a group of stuff, they'll get it right. To take away a group of stuff, you take away each part of the group, in turn, like this:

$$jewels - (rubies + emeralds) = 125$$
$$jewels - rubies - emeralds = 125$$

We still have too many unknown variables, so let's use substitution again. Looking back at our facts from the story, we can substitute for the rubies and emeralds, leaving only the number of jewels—exactly what we're trying to find.

$$jewels - \tfrac{1}{3} \text{ of } jewels - \tfrac{1}{9} \text{ of } jewels = 125$$

Finally, we're ready to do some calculations, using what we know about fractions. Starting from the left and reading the problem like we'd read a sentence, we subtract:

$$(jewels - \tfrac{1}{3} \text{ of } jewels) - \tfrac{1}{9} \text{ of } jewels = 125$$
$$\tfrac{2}{3} \text{ of } jewels - \tfrac{1}{9} \text{ of } jewels = 125$$
$$(\tfrac{6}{9} - \tfrac{1}{9}) \text{ of } jewels = 125$$
$$\tfrac{5}{9} \text{ of } jewels = 125$$

Then, if we know how much $\tfrac{5}{9}$ of something is, it's not hard to find $\tfrac{1}{9}$ of that thing. And once we know $\tfrac{1}{9}$, it's easy to figure out the whole amount.

$$\tfrac{1}{9} \text{ of } jewels = 125 \div 5 = 25$$
$$jewels = 25 \times 9 = 225$$

Do we have our answer? In all that math, there were several opportunities to make a mistake. Let's check that this number really fits our story:

$$jewels = 225$$
$$\tfrac{1}{3} \text{ of } jewels = rubies = 225 \div 3 = 75$$
$$\tfrac{1}{9} \text{ of } jewels = emeralds = 225 \div 9 = 25$$
$$\tfrac{1}{5} \text{ of } [jewels - (rubies + emeralds)] = diamonds$$
$$\tfrac{1}{5} \text{ of } [225 - (75 + 25)] = diamonds$$
$$\tfrac{1}{5} \text{ of } [125] = 25$$

Which is exactly the number of diamonds there should be, so our math checks out.

King Arthur had 225 jewels in the small chest.

[8]

> *King Arthur and Merlin went shopping at the Camelot Fair.*
> *They both had the same number of coins. Then Arthur spent 50*
> *coins on new tapestries for the walls in his castle, to keep down*
> *the winter chill. And Merlin spent ⅓ of his money on new*
> *books and scrolls to study.*
>
> *When they compared their purses at the end of the day, the*
> *ratio of King Arthur's money to Merlin's was 5:4. How many*
> *coins did they each have at the beginning of the day?*

Do my worked-out word algebra solutions seem tedious to you? It often takes longer to explain the reasoning behind the steps of math than to say, "Do it this way." But our children will understand and retain more when they make sense of what they are doing. The efficiency of algebra comes later, after they develop a solid foundation of sense-making.

Also, while you're working, your students will begin to do some of the work mentally, so they—or you as their scribe—won't have to write as much as you see here.

For this story, since both men began the day with the same amount of money, we can use the word *coins* to represent that quantity. Then we need the amount of money each man had at the end of the story. Our story gives us these three facts:

$$Arthur's\ money = coins - 50$$
$$Merlin's\ money = coins - ⅓\ of\ coins = ⅔\ of\ coins$$
$$Arthur's\ money : Merlin's\ money = 5:4$$

And we want to find the amount of money each man started with, so:

$$coins = ?$$

What can we do with this information? How can we apply common sense to this situation?

As we've seen in earlier chapters, bar model diagrams help ratio problems make sense. We could use the ratio of *Arthur's money* to *Merlin's money* to set up a bar model: five units in Arthur's bar, compared to four units for Merlin. Then we'd add a longer bar for the coins, since both men spent some of their money, and start looking for relationships between the bars.

But in this transition to standard algebra, we're trying to work directly with the abstract equations, not using the visual algebra of the bars. Without the visual model, what kind of math will make sense here?

Let's start with that ratio. You might remember from the *Romeo and Juliet* story in Chapter 7 that the ratio 5:4 means that at the end of the day, Arthur's money is ⁵⁄₄ of Merlin's money. But if you don't see that connection, how might you figure it out?

We know that ratio relationships can also be written as fractions, which makes them easier to use in calculations.

$$\text{Arthur's money} / \text{Merlin's money} = \tfrac{5}{4}$$

If we had numbers on the left-hand side of the equation, we could tell students to think of the proportion as equivalent fractions. If they figure out what factor one fraction was multiplied by to make the equivalent other fraction, they can solve the equation.

But what can students do with a fraction in word algebra?

Perhaps it would help to write that fraction as division:

$$\textit{Arthur's money} \div \textit{Merlin's money} = \tfrac{5}{4}$$

Aha! That's something we can work with. To undo division, we use inverse operations and keep it balanced.

$$(\textit{Arthur's money} \div \textit{Merlin's money}) \times \textit{Merlin's money}$$
$$= \tfrac{5}{4} \times \textit{Merlin's money}$$

so:

$$\textit{Arthur's money} = \tfrac{5}{4} \times \textit{Merlin's money}$$

So now, the ratio makes sense: Arthur ended up with more money than Merlin had, $\frac{5}{4}$ times as much. We're ready to bring in our other two facts, using substitution.

Arthur's money = coins − 50
Merlin's money = $\frac{2}{3}$ of coins
coins − 50 = $\frac{5}{4}$ × ($\frac{2}{3}$ of coins)

Finally, when we get down to an equation with a single variable, we are ready to calculate the answer. Which means, in this case, multiplying fractions.

You may wonder why I carried the word "of" through all the steps up to this point. That's because most students can reason more intuitively about a fraction *of* a quantity than about a fraction *times* something. But now that we need to calculate, we apply a keyword translation: "*of* means multiply" when working with fraction and percent problems.

coins − 50 = $\frac{5}{4}$ × ($\frac{2}{3}$ × coins)

Does this look like a Distributive Property situation to you? Many students will see it that way, which is why we must teach our kids to reason about what their equation means. The Distributive Property tells us *how multiplication interacts with addition*. It only applies when we're trying to multiply a group of separate things like the rubies and emeralds in our previous example.

Is "($\frac{2}{3}$ × coins)" a group of two different things added together, or a way to explain one single thing? Is the $\frac{2}{3}$ a separate quantity from the *coins*?

The phrase describes the amount of money Merlin had

in his purse at the end of the day, which is a single quantity. These parentheses are *not* holding a group together. They are merely a thinking aid, which means that removing them won't change the equation.

$$coins - 50 = \tfrac{5}{4} \times \tfrac{2}{3} \times coins$$

By the time students are making the transition to algebra, they should know how to multiply fractions.

$$\tfrac{5}{4} \times \tfrac{2}{3} = \tfrac{10}{12} = \tfrac{5}{6}$$
$$coins - 50 = \tfrac{5}{6} \times coins$$

When we start with any quantity and end up with $\tfrac{5}{6}$ of that quantity, then the amount we removed must have been $\tfrac{1}{6}$ of the total.

$$50 = \tfrac{1}{6} \times coins$$
$$coins = 50 \times 6 = 300$$

Merlin and King Arthur each began the day with 300 coins.

Make Your Own Math

I've numbered these prompts to match the *Word Problems Student Workbook*, which combines the problems above with a few from the next two chapters. In addition to the story prompts, you might try this assignment: Go back over the math stories you've written and choose a favorite one. Can you show how to solve it using word algebra?

[10]

Write a multistep problem set in the world of a book or movie you enjoy. Have three or more parts to compare within your story.

The numbers may include the total amount and the sizes of the parts. Or instead of telling the size of a part directly, you might use a comparison (especially a fraction or ratio) to make a multistep puzzle-within-a-problem.

How much information do you need to include so that your reader can figure it out?

[11]

Write a fraction-of-the-remainder problem set in the world of a book or movie you enjoy.

Tell a two-or-more-step problem that begins with an initial condition, then something changed, and there's a part left over, so your character works with some fraction of that remaining part.

What questions will you ask? How much information do you need to include so that your reader can figure it out?

[12]

Write a problem that uses percents. Where in your story world would a character work with percent calculations? Think about sharing out a treasure, shopping with sales or discounts, paying taxes, saving or loaning money, and things like that.

The numbers in a percent problem may include whatever counts as the base (100%), the part of that thing we are interested in, and the percentage value. Or you might use the complement, which is the percentage value for the part we are *not* interested in.

Tell two of the percentage (or the complement), the part, and the base. Then ask your reader to figure out the missing number.

[13]

Write a problem that uses both fractions and percents.

When might your character need to find a percentage of a fraction, or a fraction of a percentage? What kind of situation would lead to such a calculation?

[14]

Write a percent comparison problem.

　Why might your character be comparing two (or more) things and notice that one of them is some percentage more or less or greater or fewer than the other? What questions can you ask about such a situation?

[15]

Write any kind of problem you like.

Write Your Own Word Problems from Literature

*If you wish
to learn swimming
you have to go into the water ,
and if you wish
to become a problem solver
you have to solve problems.*

—GEORGE PÓLYA

The Story Problem Challenge

What do you get when you cross a library book or favorite movie with a math worksheet? A great alternative to math homework. The "Story Problem Challenge" forces students to consider the real-world meaning of each calculation.

The rules are simple:

(1) Start with a target calculation. Or start with an interesting bar diagram.

(2) Draw the diagram to match your calculation. Or write the calculation to match your diagram. Solve the calculation or diagram.

(3) Consider where these numbers could make sense in your book or movie universe. How might the characters use math? What sort of things would they count or measure?

Do they use money? Do they build things, or cook meals, or make crafts? Do they need to keep track of how far they have traveled? Or how long it takes to get there?

(4) Write your story problem.

The next time your children face a worksheet full of practice exercises, have them write word problems instead. Let each child choose a favorite literary universe and pick two or three calculations to convert. You do the same. Then everyone can share their favorite problems for the others to solve.

To make the game easier, players may change the numbers to make a more realistic problem, but they must keep the same type of calculation. For example, if your original calculation was 18 ÷ 3, you could change it to 18 ÷ 6 or 24 ÷ 3 or even 119 ÷ 17 to fit your story, but you can't make it 18 – 3.

Word Problems Inspired by *The Lord of the Rings*

Several online friends were studying *Elementary Mathematics for Teachers* by Thomas H. Parker and Scott J. Baldridge. We'd just finished a lesson on fraction division, and our homework challenged us to create word problems for several calculations. I decided to play in a fantasy world like that of J. R. R. Tolkien's *The Lord of the Rings* trilogy.

The assigned calculations were:

- ♦ Show 135 ÷ 5 as partitive division.
- ♦ Show 72 ÷ ⅗ as partitive division.
- ♦ Show ⅗ ÷ ⅓ as partitive division.
- ♦ Bar diagram given: 450, cut into thirds, and then two-thirds of that cut into fifths; how much are three of those fifths?
- ♦ Show 32 ÷ 3¾ as measurement division.

Partitive division means sharing. Split the whole quantity into the given number of equal parts, and then figure out the size of each part. *Measurement division* means portioning. Measure out the whole quantity into chunks of the given size, and then see how many pieces you get. (Measurement division is sometimes called *quotative division*.)

Remember our basic multiplication/division relationship from Chapter 3: The number of units times the size of each unit equals the total product.

Partitive division says, "We have this many units. What size is each unit? This many *of what* makes our total?"

Measurement division says, "We want units of this size. *How many* units can we make?"

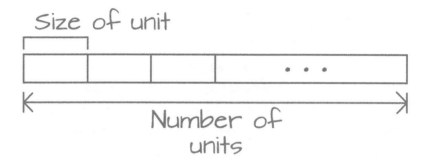

135 ÷ 5 as Partitive Division

Partitive division means sharing. This problem asks, "If we share 135 out as 5 parts, what size are the parts?" Or more simply, "5 *of what* is 135?"

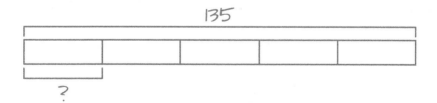

$$5 \text{ units} = 135$$
$$1 \text{ unit} = 135 \div 5 = 27$$

I need to find something that can split into five groups of twenty-seven. Since everything is whole numbers, I could use people or animals, if I want. And if I'm weaving my problems into a story, then I need something that fits near the beginning for this first puzzle.

Five ... hobbits? Yes, and naturally, they will be eating:

> *Our halfling hero and his four friends (everyone always forgets the friend who stayed behind) enjoyed one last meal before setting off on their quest. If the halflings shared equally all 135 sausages in the larder, how many sausages did each one eat?*

Do the numbers make sense? From my answer above, each halfling eats twenty-seven sausages. That makes a very full tummy, but if anyone could do it, it would be a hobbit. I could adjust the numbers down, but I prefer the story as is.

72 ÷ ⅗ as Partitive Division

Partitive division means sharing. This problem asks, "If we share 72 out as ⅗ of a part, what size are the parts?" Or more simply, "*⅗ of what* is 72?"

Notice how this problem is similar to, and different from, the whole-number division problem above. When dividing by a fraction, we find ourselves in an incredibly anti-intuitive situation. Because our total amount is only a fraction of a part, the whole shared-out "part" must be bigger than the amount we started with.

$$3 \text{ units} = 72$$
$$1 \text{ unit} = 72 \div 3 = 24$$
$$5 \text{ units} = 24 \times 5 = 120$$

So I need a story about a tad over a hundred of something and some reason to sort it into smaller parts. Since each part is a whole number, I could use people or animals.

If I were creating a worksheet, I would want to make each problem slightly more challenging than the one before. Since this problem is more difficult than the last one, it should come later in the story—maybe when our heroes got trapped in the tree or were staying with the elves.

But I like the old wizard and his fireworks. So I'm going to tell the story out of order.

The traveling wizard loaded so many fireworks into his cart that he only had time to shoot off ⅗ of them. If he blasted 72 fireworks during the party, how many did he bring in his cart?

Do the numbers make sense? From my answer above, the wizard brought a hundred and twenty fireworks. That seems reasonable.

⅗ ÷ ⅓ as Partitive Division

Partitive division means sharing. This problem asks, "If we share ⅗ out as ⅓ of a part, what size are the parts?" Or more simply, "⅓ *of what* is ⅗?"

Again, our total amount is only a fraction of a part, so the whole shared-out "part" must be bigger than the amount we started with.

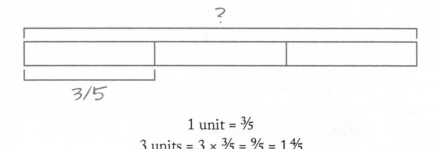

$$1 \text{ unit} = \tfrac{3}{5}$$
$$3 \text{ units} = 3 \times \tfrac{3}{5} = \tfrac{9}{5} = 1\tfrac{4}{5}$$

I need a story that describes a bit less than two of something, with a reason to split it into three parts. And getting back to the proper order of my story, I need something that happens after the hobbits leave on their quest.

I'm going to tweak the problem. I'll keep the assigned calculation, but nest it inside a larger puzzle.

> *The four halflings shared two loaves of bread for lunch on their journey. The leader wasn't very hungry, so he took a small piece and then split the rest evenly among his friends. If the other three halflings each received ⅗ of a loaf, what size piece did their leader eat?*

Do the numbers make sense? From my answer above, the three hobbits share 1⅘ loaves, and the leader eats the remaining fifth. Not much of a lunch, but adventurers can't be choosers.

Bar Diagram Given: 450, Cut into Thirds

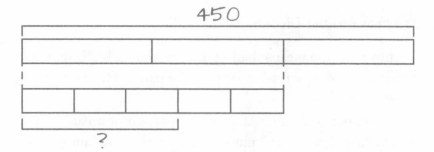

First, I solve the diagram, so I'll know what numbers to use in my story. Looking at the top bar:

3 large units = 450
1 large unit = 450 ÷ 3 = 150
2 large units = 150 × 2 = 300

Then on the remaining-part bar:

5 small units = 300
1 small unit = 300 ÷ 5 = 60
3 small units = 60 × 3 = 180

I need to create a story with numbers in the hundreds. Since everything is whole numbers, I could use people or animals, if I want. And I need to think of a reason to sort whatever-it-is into smaller parts.

How about orcs? I'm jumping far ahead in the story, but I really like this vignette.

The day before the climactic battle at the gate, a company of 450 orcs camped among the necromancer's great host. But an argument broke out over dinner, and ⅓ of them were killed.

Then ⅖ of the remainder died when a drunken troll stumbled through their camp in the dark. How many of the orcs survived to join the morning's battle?

Do the numbers make sense? From my answer above, only one hundred and eighty orcs lived through the night. Being an evil sorcerer's minion is bad for your health.

32 ÷ 3¾ as Measurement Division

Measurement division means to portion out units of a given size. This problem asks, "32 can be split into *how many pieces* that measure 3¾ each?"

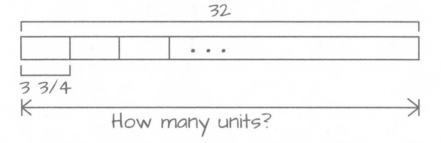

When working with fractions and mixed numbers, measurement division feels more intuitive to most people than the partitive division we did earlier. Life often gives us fractional portions to measure, so even though the mixed number makes the calculation more difficult, students find it easier to form a mental image of the situation.

$$\text{whole amount} = 32$$
$$1 \text{ unit} = 3\tfrac{3}{4}$$
$$\text{number of units} = 32 \div 3\tfrac{3}{4}$$
$$= 32 \div \tfrac{15}{4} = 32 \times \tfrac{4}{15}$$
$$= \tfrac{128}{15} = 8\tfrac{8}{15}$$

So I need to create a story with thirty-two of something inanimate, measured out into chunks of not-quite-four, and ending up with eight-and-a-fraction pieces. I can't use people or animals because the fractional parts would be too gory.

But how can it make sense for the story to end with ⁹⁄15 of something?

One might have ⁹⁄15 of a pizza, though we normally cut pizza into an even number of pieces. Maybe I could write about someone making giant pizzas (for the cave trolls?) that needed 3 ¾ pound of cheese per pizza. But why would anyone make ⁹⁄15 of a pizza? You might *eat* ⁹⁄15, but you only *cook* whole pizzas.

But when I turn it into an "eating pizza" problem, I run into another difficulty.

Each cave troll will eat 3 ¾ pizzas. If we have 32 pizzas for dinner, how many trolls can we feed?

Do the numbers make sense? From my answer above, we can feed 8 ⁹⁄15 cave trolls. But how does one feed a fraction of a troll? I don't want to be the waiter serving that ninth troll when he realizes he's still hungry.

Fail.

Try again.

Hmm… The number nine in the problem above reminds me of Tolkien's Ringwraiths. How about the following problem?

The necromancer's tailor has 32 yards of black cloth to make traveling capes for his ghostly servants. The tailor needs 3 ¾ yards for each cape. How many capes can he make?

But the fractional answer still doesn't seem to make sense. The tailor would just make eight capes and realize he didn't have enough yardage for the next one. He wouldn't make ⁹⁄15 of a cape.

Fail.

Now what?

Aha! I can tweak the problem. If I add a second question, the story is good to go.

The necromancer's tailor has 32 yards of black cloth to make

traveling capes for his ghostly servants. The tailor needs 3 ¾ yards for each cape. How many capes can he make? How much additional cloth will he need to make all 9 capes?

Do the numbers make sense? The tailor can make eight capes with the fabric on hand. And then he'll need a bit extra to finish the ninth cape. The word problem works.

It even makes a great challenge problem, since students may think the fraction $\frac{8}{15}$ measures the yardage of fabric left after making eight capes. But that fraction doesn't measure the fabric in yards—it's the fraction *of a cape*. So it shows the tailor has just over half the fabric he needs for the last cape. Tricky!

One whole cape takes 3 ¾ yards of fabric. The tailor has $\frac{8}{15}$ of that amount, so he needs $\frac{7}{15}$ more.

$$\frac{7}{15} \text{ of } 3\tfrac{3}{4} = \frac{7}{15} \times 3\tfrac{3}{4}$$
$$= \frac{7}{15} \times \frac{15}{4} = \frac{7}{4} = 1\tfrac{3}{4}$$

The tailor needs 1 ¾ yards more fabric to finish the cape.

Things To Consider in Creating Word Problems

Some quantities are discrete and countable, such as hobbits or fireworks. Other quantities are continuous, such as a loaf of bread or a length of fabric. We want our children to feel comfortable working with problems of both types.

We often think of addition and subtraction in stories that involve movement: joining or leaving, giving or taking, growth or shrinkage. But we can also write stories about comparison —how much larger or smaller (lighter or heavier, more or less expensive, etc.) is this thing than the other one.

Or we can use classification, sorting a collection of people or items into different categories. Students often find story problems without movement more difficult than the same calculation described with movement.

We often think of multiplication and division as counting or splitting groups of items. But we can also write stories with these math processes as comparison, growth, or shrinkage (how many times as much, or what fraction of the size) or as this-per-that relationships (sausages per hobbit, yards of cloth per cape).

Division of continuous quantities (sharing wine or cutting fabric) may lead naturally to fractions or decimal amounts. Remember to consider both *partitive* (sharing) and *measurement* (portioning) division in your stories.

For a more difficult puzzle, combine two or more of these problem situations into a single story.

Problem Situations for Addition and Subtraction

- ◆ The story gives the parts. Find the whole amount.

- ◆ The story gives the whole and one (or more) of the parts. Find the missing part.

- ◆ The story tells one of the parts and the difference between that and the other part. Find the missing part and/or the whole amount.

- ◆ The story tells the whole and one part. Find the difference between the parts.

- ◆ The story tells the whole amount and the difference between the parts. Find the parts.

Problem Situations for Multiplication and Division

♦ The story says that one part is some number of times greater than the other part, and it gives either the whole amount or the size of one of the parts. Find the missing information.

♦ The story has a whole thing made of some number of equal parts. Find the missing piece of information: the whole amount, the size of the part, or the number of parts.

♦ In the story, one part is a given multiple of the other part, and the story tells the whole amount. Find either part, or both of them, or the difference between them.

♦ In the story, one part is a given multiple of the other part, and the story tells the difference between the parts. Find either part, or both of them, or the whole amount.

Problem Situations with Fractions

♦ In the story, one part is some fraction of the whole amount. Two of the quantities are given: the fraction and either the part or the whole. Find the missing quantity.

♦ In the story, one part is some fraction of another part. Two of the quantities are given: the fraction and one of the parts. Find the other part, the whole amount, and/or the difference between the parts.

♦ In the story, one part is some fraction of another part. The story tells the fraction and the whole amount. Find the two parts.

Problem Situations with Ratios

♦ The story tells the ratio relationship between two (or more) parts and also gives one part or the whole amount. Find the missing parts and/or the total amount.

♦ The story tells the ratio relationship between two parts and also gives one part or the difference between the parts. Find the missing part and/or the total amount.

♦ The story tells the ratio relationship between two parts and also gives either the whole amount or the difference between the parts. Find the size of the parts.

Problem Situations with Percents

♦ In the story, one part is some percentage of the whole amount (the base). Two of the quantities are given: the percentage, the part, and/or the base. Find the missing quantity.

♦ In the story, one part is some percentage of the whole amount (the base). Two of the quantities are given: the percentage, the part, and/or the base. Find the *other* part of the whole.

♦ In the story, one part is some percentage of another part (the base). The whole amount is *not* the base. Two of the quantities are given: the percentage and either part. Find the other part, the whole amount, and/or the difference between the parts.

♦ In the story, one part is some percentage of another part (the base). The whole amount is *not* the base.

The story tells the percentage and the whole amount. Find the two parts.

♦ In a percent comparison story, one amount is some percentage greater or less than another amount (the base). Two of the quantities are given: the smaller amount, the larger amount, the difference between them, and the percentage. Find either (or both) of the other two.

A Reminder about Copyright and Trademarks

Old books are in the public domain, so anyone can use characters like Robin Hood, Romeo and Juliet, or Winnie-the-Pooh (but not the newer Disney version with the red jacket). But most books and movies are the protected intellectual property of their authors or estates, or of the company which bought those rights.

When you and your students write problems for your own private use, feel free to use your favorite characters from any story. But if you decide to share your creation beyond your own home or classroom, then be sure to "genericize" it first. Change or remove the proper names, using general descriptions instead.

For example, if your children love the Harry Potter series, they might want to use Harry or Hermione in their story problems.

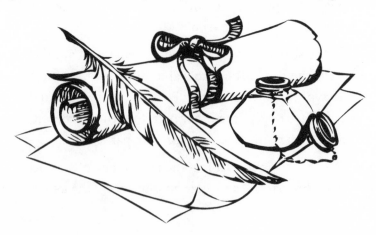

Instead, encourage them to write about "the boy wizard destined to fight an evil sorcerer." Or "the bright young witch who can master any spell."

We'd love to add your child's story to the Student Math Makers Gallery.[†]

[†] *tabletopacademy.net/math-makers*

Conclusion

There are two ways
to do great mathematics.
The first is to be
smarter than everybody else.
The second way is to be
stupider than everybody else,
but persistent.

—RAOUL BOTT

10

Making Sense of Word Problems

IN HELPING OUR CHILDREN LEARN math, word problems are much more important than straight-calculation assignments.

Plain calculations are not really problems at all, merely abstract exercises, the dead answers to some forgotten problem. The true work of mathematical thinking happens when students translate a real or imagined situation into a mathematical expression or equation. Each new problem they solve provides a little more experience of number relationships in action, and it takes plenty of these puzzles to build a deep foundation in math.

For a strong foundation, focus on making sense of each problem, reasoning through each situation, exploring the web of mathematical connections.

Give your children as much help with word problems as they truly need, but no more. Often, sitting patiently next to the child as he or she thinks through a problem is support enough. For young children, you may read the problems aloud and do the writing as they dictate. And encourage them to master problem-solving tools like working backwards, making a list or chart, looking for patterns, and drawing diagrams.

One way to help children think about the story situations is to cover up or erase the numbers. Numberless word problems force children to think about the story dynamics and figure out how the quantities involved must relate to each other. Or read the problem but leave off the question at the end. Ask children what they notice about the story. If they were writing the problem, what questions might they ask?

Bar model diagrams encourage the same sort of creative reasoning. To draw a diagram, students must decide how the quantities are related. Which is the larger amount? By how much? What parts make up the whole thing?

In the beginning, drawing a bar model takes up more space and requires more pencil-to-the-paper effort from the student than other approaches to solving simple problems. Many children solve early word problems by intuition, which makes drawing a picture seem tedious. But as word problems become more involved, bar models offer significant help for students who struggle with questions like "Do I add or multiply?" A diagram makes visible the abstract relationships between quantities, enabling the student to decide which arithmetical operation makes sense.

In upper-elementary textbooks, math problems take a leap up on the difficulty scale. Students are more mature, able to read and follow more complicated stories. Learning to justify the solution to a multistep word problem offers an early taste of mathematical proofs.

One benefit of mastering bar model diagrams is that they help children understand fractions, a topic that will haunt students in problems of ever-increasing complexity throughout middle school. Multiplication and division problems will also grow more challenging, until they become the dreaded proportional equations of algebra. These topics are notoriously difficult for students, but bar models and ratio tables provide a better foundation for understanding than any other method I've seen.

Bar Model Diagrams Are Not the Goal

As much as I praise them, bar diagrams are not our goal. In teaching elementary and middle school math, we want to move our children step by step toward the goal of understanding algebra.

Bar model diagrams are a tool that can help bridge the way to algebra. The models make visible many abstract relationships that give students trouble, such as inverse operations and fractions. But other methods are equally valid, providing options for students who don't like the visual models.

Don't memorize a specific series of steps for drawing model diagrams. I've seen lesson books that treat bar models like a magic spell or incantation that will do all the work for you—if only you follow the proper steps. That sort of rote approach will not help your children make sense of math.

Students need to think about the relationships within the problem, reasoning out for themselves how the numbers connect, forging their own path to a solution, discovering new insights along the way.

In every situation, the basic questions of mathematical problem-solving remain the same:

[1] What Do I Know?

Read the problem and understand the facts of the story. What information does the problem give you? What do you notice about the situation?

[2] What Do I Want?

Determine what you are trying to find. What question is the problem asking? What will your answer look like?

[3] What Can I Do?

Play with the puzzle. Combine different bits of information. Try making a sketch, chart, or graph, or looking for a pattern. Can you find a way to move closer to your goal?

[4] Does It Make Sense?

When you get an answer, look back at the original problem. Did you forget anything? Can you think of any way to confirm your answer? Is there another way to figure it out?

Will you be able to solve problems like this in the future? What would you say was the key step or insight in your solution?

Problem-Solving Is Applied Common Sense

Math is all about reasoning. There's no universal magic technique to solving math problems. But if your children work through the four questions above, and if they persevere, any school textbook problem will be within their grasp.

So let's play with one more math challenge. Here's a two-part problem that algebra or bar diagrams probably *won't* help you solve. Can you think it through?

Challenge 4: The Hyperdrive Control Panel

[a] Captain Henry wants to upgrade the old hyperdrive configuration on his starship. The power controls consist of two toggle switches on his dashboard, and each switch can point either up or down.

Every arrangement of these switches signals the engine to run at a different power level. With the current configuration, the captain can apply four levels of hyperdrive power:

♦ *down, down (the lowest level)*

- *down, up*

- *up, down*

- *up, up (the highest level)*

But his engine can actually run at ten different power levels. How many additional switches should the ship's engineer install to let Captain Henry signal all ten hyperdrive levels?

[b] Chief Engineer John fixes the hyperdrive controller, putting in the new switch(es) that the captain ordered.

Then John decides it's time to upgrade the ship's engine, too. But he doesn't want to make the controls too confusing.

How many additional hyperdrive power levels will the new toggle-switch configuration handle, without installing still more switches?

No, I won't solve this one for you. Put those problem-solving questions into practice before you try teaching them to your kids. If you get completely stumped, try explaining the problem to another person. That can help your mind notice an idea it missed until you put everything into words.

Most of all, with any math problem, seek understanding. Practice the art of logical reasoning, making sense of whatever situations you meet, eager to learn new things, eager to understand more deeply the things you already know, enjoying the adventure of life.

...and may the Math be with you!

Appendixes

Online Resources

If you enjoyed the problems in this book and want to use them with your students, look for the companion *Word Problems Student Workbook* at your favorite online bookseller.

As you teach children to make sense of math, avoid using catch-phrases or slogans that hide what's actually happening with the numbers. Find meaningful alternatives in Tina Cardone's free booklet *Nix the Tricks*.

nixthetricks.com

For tips on using mental math to help children develop reasoning skills, follow Pam Harriss's podcast, *Math Is Figure-Out-Able*.

podcast.mathisfigureoutable.com

For Young Children

Find numberless word problem tips at Brian Bushart's site.

numberlesswp.com

To get more practice creating bar diagrams, your students may enjoy the Thinking Blocks interactive tutorial at the Math Playground website.

mathplayground.com/thinkingblocks.html

For Older Students

Numberless word problems are not just for little kids. Check out Farrar Williams's book *Numberless Math Problems: A Modern Update of S.Y. Gillian's Classic Problems Without Figures*, available in ebook or paperback.

farrarwilliams.wordpress.com/2018/09/04/math-with-no-numbers

An Adventure in Problem-Solving

If your older students enjoyed the problems in this book, they may be ready for a real math adventure story: *The Arithmetiquities.*

> "When the world of Sfera is threatened by the machinations of a malevolent sorcerer, it will be up to a band of unlikely heroes to become the brightest light in the darkness.
>
> "The adventurers fan out across the land to find and retrieve the Arithmetiquities, a set of ancient mathemagical artifacts. Will they collect the relics and perform the arcane rites before evil engulfs the world?"

The Arithmetiquities is a fantasy adventure story for middle and high school students by former teacher Jason Ermer, told through a sequence of 36 puzzles. It's sort of like a fantasy escape room for math and logic. Great fun!
arithmetiquities.org

Problem-Solving with James Tanton

The ninja master of math education, James Tanton offers wonderful problem-solving tips for middle and high school students. For example, check out:

"How To Think like a School Math Genius (Five Principles of Problem-Solving)"
jamestanton.com/?p=1097

"Two Key—But Ignored—Steps to Solving Any Math Problem"
medium.com/q-e-d/two-key-but-ignored-steps-to-solving-any-math-problem-8cd927bf60a0

"Ten Problem-Solving Strategies"
jamestanton.com/?p=1193

"Puzzles and Other Cool Math"
jamestanton.com/?category_name=puzzles

Answers to the Student Workbook

Lay the Foundation: One-Step Problems

[1] Papa read 39 books.

[2] The boy made 32 snowballs. The children made 50 snowballs altogether.

[3] There were 38 fish left.

[4] It took 55 centimeters of ribbon to make the bow.

[5] The penguins performed 8 shows in all.

[6] Papa fell down 30 steps in all.

[7–11] Answers will vary.

Build Modeling Skills: Multistep Problems

[1] Ben and his father sold 611 candles.

[2] James sold 653 newspapers, and the brothers sold 1,483 altogether.

[3] There were 5 books per shelf.

[4] Ben read 28 more history books than science books.

[5] Ben still needs to collect 400 pounds.

[6] There were 3,001 people at the party.

[7–11] Answers will vary.

Master the Technique: From Multiplication to Fractions

[1] There were 140 creatures at the party, with 28 dancers in each circle.

[2] The professor had 218 books in his study.

[3] There were 58 black dwarfs.

[4] ⅙ of the marmalade roll remains.

[5] They used ⅝ of the pitcher of milk.

[6] The prisoner pushed the sledge 16 times.

[7] There were 50 creatures with the lion prince at his pavilion.

[8] Twelve of the creatures could not fly.

[9] They used 8½ meters of rope.

[10] There were 40 sea people performing at the coronation party.

[11–20] Answers will vary.

Extend Your Skills: Measurement and Decimals

[1] Together, the children drank 0.9 liter of barley water. Peter drank 0.5 liter more than Phyllis drank.

[2] The total length of the two pieces of string was 5.1 m.

[3] The old gentleman received £3.26 in change.

[4] The children drank 1.2 liters of juice in all.

[5] Mother had £18.20 left after buying the tickets.

[6] Each little bottle held 0.3 liter of medicine.

[7] Each meter of flannel cost £1.60.

[8] One chocolate bar weighed 0.19 kg.

[9] The total weight of all four parcels was 7.95 kg.

[10] Each packet of flour weighed 1.13 kg, or 1 kg 130 g.

[11] It took 1 h 15 min for the workmen to clear one meter of track.

[12] Jim spent 17 h 30 min working on the fence.

[13] Mother used 14.63 m of material altogether.

[14] Mother saved 5 pence per sticky bun.

[15–22] Answers will vary.

Reap the Reward: Ratios and More Fractions

[1] There are 48 pastries altogether.

[2] Charles had 38 gold pieces.

[3] The Great Goblin sent 400 soldiers to his cousin.

[4] The cave creature ate 2 ½ fish at each meal.

[5] The bear-man had ⅙ of his loaf of bread left.

[6] The butler drank ⅜ of a gallon of wine.

[7] ¹⁄₁₂ of the dragon's treasure was precious gems.

[8] There were 1,000 elves marching with their king.

[9] 240 archers followed the master bowman to battle.

[10] The halfling received 2,700 pieces of gold, 1,350 gems, and 1,950 pieces of silver.

(Solutions for problems 11–13 are given in Chapter 9: Write Your Own Word Problems from Literature.)

[11] The wizard brought 120 fireworks.

[12] Each halfling ate 27 sausages.

[13] The three halflings shared 1⅘ loaves, and their leader ate the remaining ⅕.

[14–21] Answers will vary.

Conquer Monster Topics: Percents, Rates, and Proportional Reasoning

[1] Miranda had 0.725 kg of sugar left.

[2] 1% of the flowers were love-potions.

[3] The shepherd saved 60% of the treasure for Perdita.

[4] Don Pedro wagered 24 gold pieces.

[5] There were 27 servants in the banished duke's group.

[6] Julia paid 1,390.50 L for the gift.

[7] Antonio expected to gain approximately 7,500 ducats.

[8] The spring produced 2.5 liters per minute.

[9] King Lear needed 25 days to travel to the earl's castle.

[10] The slowly simmering potion made 15 bubbles per minute.

[11] Helena took 25 days to travel that distance.

[12] The ratio of length to width to perimeter is 3:2:10.

[13] Hamlet had 60 guilders.

[14] There were 2,100 gentleman officers under Othello's command.

[15] Marina earned 150 talents for her teaching.

[16] There were 104 people at Lord Capulet's party.

[17–21] Answers will vary.

Move Toward Algebra: Challenge Problems

(Solutions for problems 1 and 2 are given in Chapter 9: Write Your Own Word Problems from Literature.)

[1] Only 180 orcs lived through the night.

[2] The tailor can make 8 capes with the fabric on hand. And then he'll need 1 ¾ yards more to finish the ninth cape.

[3] The smuggler has 50 money chips to her name.

[4] The princess sorted 160 pieces of Dragon Philip's treasure.

[5] There were 75% fewer swordsmen than archers.

[6] King Arthur was sending 12 knights on the quest.

[7] King Arthur had 225 jewels in the small chest.

[8] Merlin and King Arthur each began the day with 300 coins.

[9a] The captain needs 2 additional switches, for a total of 4 toggle switches in all.

[9b] And with these new switches, the captain can control 16 levels of hyperdrive power, which is 6 more than his current engine will handle.

[10–15] Answers will vary.

Credits

Mr. Popper's Penguins ©1938 Richard and Florence Atwater

Poor Richard ©1941 James Daugherty

The Lion, the Witch and the Wardrobe ©1949 C. S. Lewis

The Railway Children ©1905 E. Nesbit (public domain)

The Hobbit ©1937 J. R. R. Tolkien

The Lord of the Rings ©1954 J. R. R. Tolkien

Tales from Shakespeare ©1807 Charles and Mary Lamb (public domain)

Star Wars ©1977 George Lucas, 20th Century Fox

Dealing with Dragons ©1990 Patricia C. Wrede

Idylls of the King ©1859–1885 Alfred, Lord Tennyson (public domain)

"Storying—encountering the world and understanding it ..." Gordon Wells, quoted in *Writing in the Teaching and Learning of Mathematics* by J. Meier and T. Rishel.

"It is the duty of all teachers ..." Paul Halmos, quoted in *Out of the Mouths of Mathematicians* by Rosemary Schmalz.

"We teachers so often hear ...," Herb Gross quoted by Jerome Dancis in "Reading Instruction for Arithmetic Word Problems," July 17, 2007. *www2.math.umd.edu/~jnd/subhome/Reading_Instruction.htm*

"The Purpose of Word Problems," passage compiled, with permission, from two articles by Andre Toom, Ph.D.
toomandre.com/my-articles/engeduc/index.htm

"Word problems are very valuable ..." excerpted from "Word Problems in Russia and America," academic paper, November 6, 2010; extended version of a talk at the Meeting of the Swedish Mathematical Society, June 2005.
de.ufpe.br/~toom/travel/sweden05/WP-SWEDEN-NEW.pdf

"For example, coins ..." from "Word problems: Applications vs. Mental Manipulatives," *For the Learning of Mathematics,* v. 19, March 1999.
toomandre.com/my-articles/engeduc/MANIPUL.PDF

"Wholenumberville (.) Fractionland" image courtesy of Christopher Danielson, from "The Triangleman Decimal Institute [TDI]" on Overthinking My Teaching blog, Sept. 27, 2013.
*christopherdanielson.wordpress.com/2013/09/27/
the-triangleman-decimal-institute-tdi*

"If you wish to learn swimming ..." George Pólya, *Mathematical Discovery: On Understanding, Learning and Teaching Problem Solving,* Wiley, combined edition 1981.

"There are two ways ..." Raoul Bott, quoted in The MacTutor History of Mathematics archive.
www-history.mcs.st-and.ac.uk/Biographies/Bott.html

Special Thanks

THIS BOOK CAME TO PRODUCTION with the help of many wonderful people who backed the project on Kickstarter. Your generous support and encouragement keeps me going!

Special thanks to:

Algie Lane

Amber G.

Amy Campbell

Anneliese McKenna

BO

Brinkman family

Bush family

Chee Lup Wan

Childers family

Claflin family

Cook family

Cynthia Gallo

Dave Holets

Elizabeth Cousin

Erika Reid

Feliciano family

Finley & Clover Adcock

Forest Grove School

Glenda L. Johnson

Halstead family

Holly Luzzi

Iwona

Jane Miller

Janette Fletcher

Jennifer Felker

Jo Oehrlein

JoAnne Growney

Karen Reddick

Kathy Kuhl

Kell

Kim Miller

Kirk Lunde

Krystal Bohannan

Laura Cole

Laura Kaplan

Leslie L.

Libby Dilg

Matt Hargett

Morrison family

Nancy Paulson Fox

Pattie Perry

Paul Lazenby

Perrine Gilchrist

Reeves family

Reshelving Alexandria

Riley family

Santos family

Shawnda Lind

Sophia Antoinette Praeger

Sykora family

Teresa Gonczy O'Rourke

Tracy Popey

Trisha J. Wooldridge

Vic Beaumont

Vissers family

Wanda Aasen

Weiss family

William, Annabelle, Anya Lee

Yelena

About the Author

FOR MORE THAN THREE DECADES, Denise Gaskins has helped countless families conquer their fear of math through play. As a math coach and veteran homeschooling mother of five, Denise has taught or tutored at every level from preschool to precalculus. She shares math inspirations, tips, activities, and games on her blog at DeniseGaskins.com.

Denise encourages parents and teachers to look at math with fresh eyes. "We want to explore the adventure of learning math as mental play, the essence of creative problem solving. Mathematics is not just rules and rote memory. Math is a game, playing with ideas."

Get More Playful Math

Join Denise's free newsletter list to receive an eight-week "Playful Math for Families" email course with math tips, activities, games, and book excerpts. Also, about once a month, she'll send out additional ideas for playing math with your kids. And you'll be one of the first to hear about new math books, revisions, and sales or other promotions.

Math Games, Tips, and Activity Ideas for Families.

TabletopAcademy.net/MathNews

Books by Denise Gaskins

tabletopacademy.net/playful-math-books

"Denise has gathered up a treasure trove of living math resources for busy parents. If you've ever struggled to see how to make math come alive beyond your math curriculum (or if you've ever considered teaching math without a curriculum), you'll want to check out this book."
— KATE SNOW, AUTHOR OF MULTIPLICATION FACTS THAT STICK

Let's Play Math: How Families Can Learn Math Together — and Enjoy It

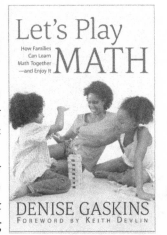

Transform your child's experience of math!

Even if you struggled with mathematics in school, you can help your children enjoy learning and prepare them for academic success.

Author Denise Gaskins makes it easy with this mixture of math games, low-prep project ideas, and inspiring coffee-chat advice from a veteran homeschooling mother of five. Drawing on more than thirty years of teaching experience, Gaskins provides helpful tips for parents with kids from preschool to high school, whether your children learn at home or attend a traditional classroom.

Don't let your children suffer from the epidemic of math anxiety. Pick up a copy of *Let's Play Math*, and start enjoying math today.

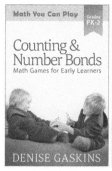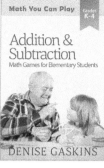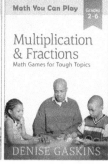

The *Math You Can Play* Series

You'll love these math games because they give your child a strong foundation for mathematical success.

By playing these games, you strengthen your child's intuitive understanding of numbers and build problem-solving strategies. Mastering a math game can be hard work. But kids do it willingly because it's fun.

Math games prevent math anxiety. Games pump up your child's mental muscles, reduce the fear of failure, and generate a positive attitude toward mathematics.

So what are you waiting for? Clear off a table, grab a deck of cards, and let's play some math.

The *Playful Math Singles* Series

The *Playful Math Singles* from Tabletop Academy Press are short, topical books featuring clear explanations and ready-to-play activities.

312 Things To Do with a Math Journal includes number play prompts, games, math art, story problems, mini-essays, geometry investigations, brainteasers, number patterns, research projects for all ages.

70+ Things To Do with a Hundred Chart shows you how to take your child on a mathematical adventure through playful, practical activities. Who knew math could be so much fun?

More titles coming soon. Watch for them at your favorite bookstore.